T0310017

Control, Optimization, and Smart Structures

Control, Optimization, and Smart Structures

High-Performance Bridges and Buildings of the Future

HOJJAT ADELI
The Ohio State University

AMGAD SALEH
Lucent Technologies

JOHN WILEY & SONS, INC.
New York / Chichester / Weinheim / Brisbane / Singapore / Toronto

Library of Congress Cataloging-in-Publication Data:

Adeli, Hojjat, 1950–
    Control, optimization, and smart structures : high-performance
bridges and buildings of the future / Hojjat Adeli, Amgad Saleh.
        p.   cm.
    ISBN 0-471-35094-X (cloth : alk. paper)
    1. Structural optimaization--Data processing. 2. Computer-aided
design. 3. Adaptive control systems. 4. Smart structures.
    5. Bridges--Design and construction. I. Saleh, Amgad. II. Title.
    TA658.8.A34 1999
    624.1--dc21   99-21925

Printed in the United States of America.

10 9 8 7 6 5 4 3 2 1

To

Nahid, Anahita, Amir Kevin, Mona, and Cyrus Dean Adeli

and

Nivine, Kareem, Omar, and Yaseen Saleh

CONTENTS

Preface

Through the use of active controllers a structure can modify its behavior during dynamic loading such as impact, wind, or earthquake loading. Such structures with self-modification capability are called adaptive or smart structures. The smart structure technology will have enormous consequences in terms of preventing loss of life and damage to structure and its content especially for large structures with hundreds of members. In this futuristic cross-disciplinary book, computational models and algorithms are presented for active control of a new generation of large adaptive structures subjected to various types of dynamic loading such as impact, wind, earthquake, and blast loading. This is achieved through ingenious integration of four different technologies with structural engineering: control theory, optimization theory, sensor/actuator technology, and high-performance computing.

An important focus of the book is simultaneous optimization of the structure and control system in order to minimize the cost of both structure and control system. The formulation of such an integrated structural/control optimization problem is complex. Its solution for large structures with hundreds of members requires high-performance computing resources. As such, employing the vectorization and parallel processing capabilities of multiprocessor supercomputers, a number of new, robust, and efficient concurrent algorithms are presented for various aspects of the solution process such as the complex eigenvalue problem of a real unsymmetrical matrix and the Riccati equation.

The computational models and high-performance concurrent algorithms are applied to two different classes of structures: bridge and multistory high-rise building structures. Extensive parametric investigations are presented on the optimal placement of controllers for both bridge and multistory building structures. With the rapid and continuous improvement of the sensor/actuator technology as well as the increase in the processing power of computers and their decreasing price trend, we envisage a new generation of high-performance structures where sensors, actuators, and microprocessors sense their environment, measure the resulting dynamic effects, and react to compensate for and reduce their destructive consequences.

Hojjat Adeli and Amgad Saleh

Acknowledgment

The work presented in this book is partially based on the work sponsored by Cray Research, Inc. Supercomputing time on the Cray YMP machine was provided by the *Ohio Supercomputer Center*. Parts of the research presented in this book were published in several journal articles in, *Computer-Aided Civil and Infrastructure Engineering* (published by Blackwell Publishers), *Journal of Aerospace Engineering* (published by American Society of Civil Engineers), *Journal of Structural Engineering* (published by American Society of Civil Engineers), *International Journal of Solids and Structures* (published by Elsevier), and *Mechatronics* (published by Elsevier), as noted in the list of references.

About the Authors

Hojjat Adeli received his Ph.D. from Stanford University in 1976. He is currently Professor of Civil and Environmental Engineering and Geodetic Science, Director of Knowledge Engineering Lab, and a member of Center for Cognitive Science at The Ohio State University. A contributor to 47 different research journals, he has authored over 300 research and scientific publications in diverse engineering and computer science disciplines. Professor Adeli has authored/co-authored eight pioneering and trend-setting books including *Parallel Processing in Structural Engineering*, Elsevier, 1993, *Machine Learning - Neural Networks, Genetic Algorithms, and Fuzzy Systems*, John Wiley, 1995, *Neurocomputing for Design Automation*, CRC Press, 1999, *Distributed Computer-Aided Engineering*, CRC Press, 1999, and *High-Performance Computing in Structural Engineering*, CRC Press, 1999. He has also edited 12 books including *Knowledge Engineering - Volume One Fundamentals*, and *Knowledge Engineering - Volume Two Applications*, McGraw-Hill, 1990, *Parallel Processing in Computational Mechanics*, Marcel Dekker, 1992, *Supercomputing in Engineering Analysis*, Marcel Dekker, 1992, and *Advances in Design Optimization*, Chapman and Hall, 1994. Professor Adeli is the Founder and Editor-in-Chief of the research journals *Computer-Aided Civil and Infrastructure Engineering* which he founded in 1986 and *Integrated Computer-Aided Engineering* which he founded in 1993. He is the recipient of numerous academic, research, and leadership awards, honors, and recognition. Most recently, he received The Ohio State University *Distinguished Scholar Award* in 1998 "*in recognition of extraordinary accomplishment in research and scholarship*". In September 1998 he was awarded a

United States Patent for his neural dynamics model for design automation and optimization. He is listed in 25 Who's Who's and archival biographical listings such as *Two Thousands Notable Americans*, *Five Hundred Leaders of Influence*, and *Two Thousands Outstanding People of the Twentieth Century*. He has been an organizer or member of advisory board of over 135 national and international conferences and a contributor to 104 conferences held in 35 different countries. He was a Keynote/Plenary Lecturer at 25 international computing conferences held in 20 different countries. Professor Adeli's research has been recognized and sponsored by government funding agencies such as the *National Science Foundation*, *Federal Highway Administration*, and *U.S. Air Force Flight Dynamics Lab*, professional organizations such as the *American Iron and Steel Institute* (AISI), the *American Institute of Steel Construction* (AISC), state agencies such as the *Ohio Department of Transportation* and the *Ohio Department of Development*, and private industry such as Cray Research Inc. and Bethlehem Steel Corporation. He is a Fellow of the *World Literary Academy* and *American Society of Civil Engineers*.

Amgad Saleh received his Ph.D. degree from The Ohio State University in 1996. He is currently with Lucent Technologies in Naperville, Illinois, working in the area of Intelligent Network Performance. Dr. Saleh is the co-author of sixteen research journal articles and papers in proceedings of conferences in the areas of control, optimization, parallel processing, continuum mechanics, and solid modeling.

Control, Optimization, and Smart Structures

CHAPTER 1

INTRODUCTION

1.1 ACTIVE CONTROL OF STRUCTURES

Active control of structures has been recognized as one of the most challenging and significant areas of research in structural engineering in recent years (Housner et al., 1996 and Kobori, 1996). A structure with active controllers can modify its behavior during dynamic loadings. Such a structure is called adaptive/smart structure.

In a smart structure we design a predetermined number of members to be actively controlled members. Each such member has an actuator. Sensors are placed at the joints to observe displacements and velocities in the directions of the predetermined degrees of freedom. Actuators apply the required forces for the appropriate correction to the uncontrolled response which is determined by signal conditioners (devices that translate the output signals from the sensors to forces to be exerted by the actuators) in the state feedback control system (Figure 1.1).

1.2 ACTUATORS

Actuators are devices that can produce given forces or strains. Actuators that produce forces in the axial direction are called linear actuators. The driving force mechanism in linear actuators can be hydraulic, electric, or electro-hydraulic. Actuator forces can range from a few newtons (pounds) to a few hundred kilo newtons (kilo pounds). In this work, we consider linear actuators only. Linear actuators are usually positioned at member ends (Figure 1.2). One actuator placed at one end of the member is usually sufficient. But, when the force exertion capacity of one actuator is not enough, two actuators may be used, one at each end of the member.

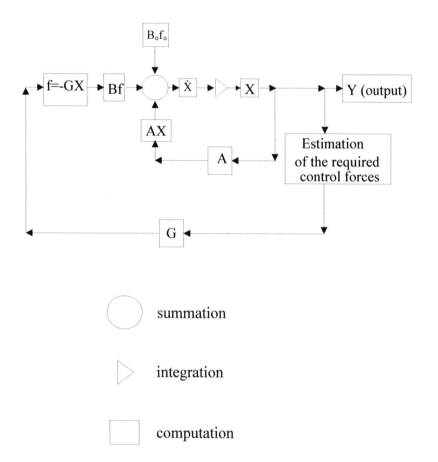

Figure 1.1: State feedback control system

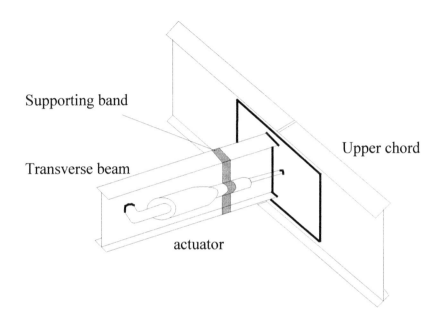

Supporting band

Upper chord

Transverse beam

actuator

**Figure 1.2: Example of positioning an actuator in a truss
bridge structure**

An example of an electric actuator is presented schematically in
Figure 1.3. It consists of three main parts: thrust unit, motor, and
accessories. The thrust unit provides linear motion and is connected
to an electric motor. Accessories are used to actuate switches and to
produce feedback signals and digital signals for computer-
controlled operations. Electric actuators can be 50-200 cm long with

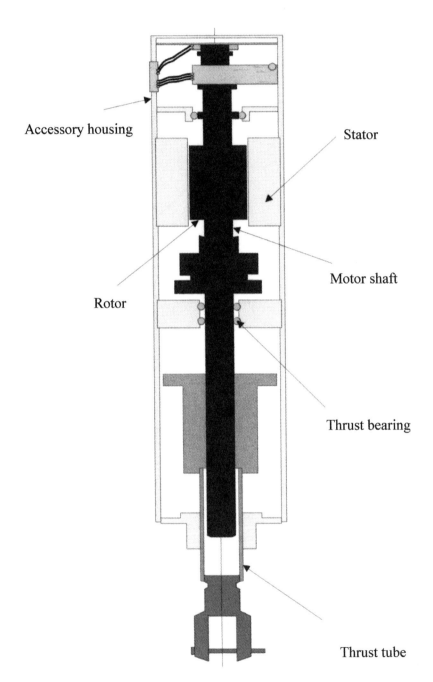

Accessory housing

Stator

Rotor

Motor shaft

Thrust bearing

Thrust tube

Figure 1.3: Schematic view of an electric linear actuator

their cross-sectional diameters ranging from 10 to 40 cm. The level of force produced by electric actuators is in the range 1 to 900 kN. In general, the size of an actuator increases with its capacity to produce the required thrust. Electro-hydraulic actuators can be as long as 360 cm with cross-sectional diameters up to 120 cm. The level of force produced by electro-hydraulic actuators can be up to 2250 kN. In this work, we consider electric actuators that can produce forces up to 200 kN, for their wide commercial availability and their appropriate dimensions for the structural applications.

1.3 OVERVIEW OF CHAPTERS

Active control of large smart structures with hundreds of members requires an inordinate amount of computer processing time and at present can be achieved within a reasonable time using the multiprocessing power of high-performance computers. As such, a good portion of this book is about development of efficient parallel algorithms on such machines. The terms parallel processing, concurrent processing, and multitasking are used interchangeably throughout this book.

Chapter 2 presents an overview of the computing environment used, and various parallel processing approaches on a shared memory multiprocessor computer including macrotasking, microtasking, and autotasking. Key issues in developing efficient parallel algorithms such as vectorization, racing condition, load balancing, and stripmining (combining vectorization with microtasking) are discussed. In order to study the most effective parallel processing approaches and stratagems three multitasking algorithms are presented for optimization of axial-force space structures (trusses) and applied to three examples ranging in size from 526 to 3126 members. Vectorization and parallel processing speedups are presented for various algorithms and important conclusions are extracted to be used in subsequent chapters.

Chapter 3 presents the formulation of the integrated structural/control optimization problem. The goal is to simultaneously optimize both the structure and the control system. The optimization approach used is the optimality criteria approach. This selection is based on its simplicity. Considering the highly complex nature of the integrated structural/control optimization problem a decision was made to use a simple optimizer in order to avoid additional requirements for computer processing power. The most time-consuming parts of the problem are the solutions of the complex eigenvalue problem and the Riccati equation encountered in the formulation.

Chapter 4 starts with reviewing various approaches for the solution of the complex eigenvalue problem including the LR method (Rutishauser, 1990), the QR method (Stewart, 1973), and the method of matrix iterations (Dongarra and Sidani, 1993). Their suitability and adaptability for development of parallel algorithms is discussed. Then, parallel algorithms are presented for the solution of the complex eigenvalue problem using the general approach of matrix iterations. Various stratagems are presented to minimize the parallel processing bottleneck and achieve maximum efficiency. The results obtained by the parallel algorithms are verified by application to two small examples reported in the literature as well as existing sequential software packages. Then, the robustness and efficiency of the parallel algorithms are demonstrated by application to four large examples including a 1632×1632 real unsymmetric matrix obtained from a 21-story space truss structure.

The Riccati equation is a class of matrix quadratic algebraic equations arising in the study of controlled dynamic systems as well as other optimal control problems. Chapter 5 starts with reviewing various approaches for the solution of the Riccati equation including the eigenvector method (Meirovitch, 1990), the Schur method (Paige and Loan, 1981), the sign function method (Gardiner and Laub, 1986), and the Newton's iterative method (Sandell, 1974). Their suitability and adaptability for development of parallel algorithms is investigated. Then, parallel algorithms are presented for the solution of the Riccati equation employing the eigenvector approach. A major portion of the computation time is consumed by the solution of the complex eigenvalue problems. As such, the algorithms presented in this chapter use the parallel algorithms presented in the previous chapter. The accuracy of the results obtained by the parallel algorithms is verified by application to two small examples reported in the recent literature, a two-bar truss and a 12-bar space truss. Then, the robustness and efficiency of the parallel algorithms are demonstrated by application to three large examples resulting from a two-span continuous truss bridge structure with 388 members, a 21-story space truss structure with 205 members, and a 12-story moment-resisting space frame with 152 members.

Building upon the materials presented in previous chapters, Chapter 6 presents a computational model for active control of large structures subjected to dynamic loading such as impact, wind, or earthquake loading. A robust parallel algorithm is presented for the recursive solution of the open-loop (uncontrolled structure) and closed-loop (controlled structure) systems. The computational model is applied to active control of smart bridge structures. Various schemes for placement of controllers in bridge structures are investigated. Results are presented for three example bridge structures: a single-span truss bridge, a two-span truss bridge, and a one-span curved truss bridge. Recommendations are made on the optimal placement of controllers.

Chapter 7 presents the application of the computational model and parallel algorithms described in Chapters 4 to 6 to multistory building structures. Various schemes for placement of controllers in multistory building structures are investigated. Three examples are presented: a 7-story space moment-resisting frame with a cloverleaf plan, a 7-story space moment-resisting frame with setbacks, and a 12-story space moment-resisting frame with and without bracings.

Chapter 8 presents the application of the computational model and parallel algorithms to multistory building structures subjected to blast and bomb loadings. Both internal blast loading at different floor levels and external blast loading from outside the structure are considered for the three example structures presented in Chapter 7.

Building upon the materials presented in previous chapters, Chapter 9 picks up the problem of simultaneous optimization of the control system and structure which should be the ultimate goal. Outline of a parallel algorithm for solution of this highly complicated and CPU-intensive problem is presented. Two example steel bridge and a multistory moment-resisting space steel frame structures are presented. A discussion on memory and CPU requirements sheds lights on the size and complexity of the problem being solved. The chapter ends with concluding remarks.

CHAPTER 2

MICROTASKING, MACROTASKING, AND AUTOTASKING

2.1 INTRODUCTION

Since their invention in the 1940s, computers have been built around one basic plan: a single processor, connected to a single store of memory, executing one instruction at a time. Since the beginning of 1990s, it has been asserted that this model would not be dominant much longer. Parallel processing, the method of having many small tasks solve one large problem by running the small tasks concurrently on many processors, has emerged as a key enabling technology in modern computing (Adeli, 1992a&b, and Adeli and Kamal, 1993).

Two different architectures are used in parallel processing (Adeli, 1992a). The first architecture is the shared-memory machines where several processors are combined in one machine with a large shared memory. The second architecture is the distributed-memory system, where a number of processors with their own local memory are connected by a network (Adeli and Kumar, 1999). Recently in some cases, the two aforementioned architectures have been combined together to produce an unmatched computational power (Lindmann, 1998).

Our goal is to develop efficient algorithms employing both vector processing and multitasking capabilities of multiprocessor supercomputers (Adeli and Soegiarso, 1999). In this chapter we explore judicious combination of various multitasking approaches, that is, microtasking, macrotasking, and autotasking with the goal of achieving a vectorized and multitasked algorithm for optimization of large structures with maximum speedup performance. In both analysis and optimization of large structures a significant portion of processing time is used for solution of simultaneous linear equations. Two different storage schemes are discussed for solution of simultaneous equations using the Gauss elimination method: the banded storage approach and the column-active skyline approach (Bathe, 1982).

2.2 COMPUTING ENVIRONMENT

The computer used in this work is the Cray YMP8E/128. It is a shared-memory machine with eight processors, up to 32 Mwords of main memory, dual instruction mode for 32-bit addressing, multiple memory ports, and a 6-ns clock cycle. It supports vectorization and multitasking in FORTRAN and C computer languages, using the UNICOS operating system. UNICOS is derived from the AT&T UNIX System V operating system and is also based in part on the Fourth Berkeley Software (CRAY, 1990).

The computer language used in this work is the CRAY Standard C version 3.0 (CRAY, 1990). This is the first version of C that supports macrotasking, microtasking, and autotasking on the CRAY supercomputer machine. It also allows macrotasking and microtasking to be combined. But, it does not allow combination of autotasking with either macrotasking or microtasking.

2.3 VECTORIZATION

On Cray YMP8E/128 a single vector operation can produce a vector containing up to 64 values. Vectorization is performed on the innermost nested loops. The code segments may have to be rearranged in order to optimize the vectorization performance. Some complications in the loop structure may prevent loop vectorization (Adeli and Soegiarso, 1999).

2.4 MULTITASKING

Concurrent processing on Cray YMP8E/128 is performed by microtasking, macrotasking, and autotasking.

2.4.1 Microtasking

Microtasking is parallel processing at the loop level. It is implemented by inserting compiler directives. Existing serial codes can be rather easily microtasked without creating new concurrent algorithms. But, in most cases and for computation intensive jobs microtasking by itself does not yield high speedup; it should be combined with macrotasking in order to achieve maximum performance.

2.4.2 Macrotasking

Macrotasking is performed at function level. Normally, major tasks that can be processed concurrently are macrotasked. Macrotasking is implemented by function calls and is suitable for tasks requiring large processing time because its overhead is large compared with that of microtasking. Macrotasked tasks should be identified when the general concurrent algorithm is developed.

2.4.2 Autotasking

Autotasking is the automatic distribution of tasks to multiple processors by compiler. It attempts to detect parallelism in the code at the loop level automatically.

2.5 CONCURRENT PROCESSING ISSUES

2.5.1 Racing Condition

In a shared-memory machine, a racing condition exists whenever more than one processor tries to access the same memory location at the same time. This may cause erroneous results (Adeli and Kamal, 1993). In structural analysis and optimization problems a racing condition is encountered when the structure stiffness matrix is assembled.

The problem of racing condition is handled by defining *guarded regions*. A guarded region is a region of code in a program that can not be run concurrently. Only one processor at a time is allowed to enter a guarded region. Thus, erroneous results due to racing condition are avoided.

2.5.2 Load Balancing

Good load balancing is defined as even (or almost even) distribution of work among processors. This is achieved by dividing iterations or tasks into a number of chunks of equal size, equal to the number of processors used. Each chunk consists of a number of consecutive iterations or tasks. For assembling the structure stiffness matrix, we divide the structure into a number of substructures, equal to the number of processors used (Adeli and Kamal, 1993). Each substructure has roughly equal number of members. Each processor calculates the element stiffness matrices of a substructure and assembles them into the structure stiffness matrix. Similarly, for assembling the structure load vector, we divide the structure into a number of substructures, equal to the number of processors used. But, in this case each substructure has roughly equal number of nodes. Each processor assembles the nodal forces in a substructure into the structure load vector. For solution of the simultaneous linear equations, we distribute the operations during different phases of the solution among the processors such that the dependency between the equations is avoided as much as possible.

2.5.3 Combining Vectorization with Microtasking (Stripmining)

Stripmining refers to combination of vectorization with microtasking at loop levels (CRAY, 1990). One must partition long jobs into a set of nested loops so that the most inner loops lend themselves to vectorization and the outer loops are processed by microtasking. Stripmining has the potential of reducing the wall-clock time (the actual clock time) substantially especially for large structures, where the overhead due to microtasking is small compared with the reduction in execution time due to stripmining. To achieve maximum gain from stripmining the tasks must be carefully distributed among processors (load balancing).

2.6 STRUCTURAL OPTIMIZATION

In this section we introduce briefly the problem of structural optimization using the optimality criteria approach. The discussion is general but formulation is limited to axial-load truss structures. This problem is defined as:

Minimize:

$$W(\mathbf{x}) = \sum_{m=1}^{M} \rho_m l_m x_m \tag{2.1}$$

subject to

$$d_i^L \le d_{ig}(\mathbf{x}) \le d_i^U; \qquad i = 1,2,\ldots,I; \quad g = 1,2,\ldots,G \qquad (2.2)$$

$$\sigma_m^L \le \sigma_{mg}(\mathbf{x}) \le \sigma_m^U; \qquad m = 1,2,\ldots,M; \quad g = 1,2,\ldots,G \qquad (2.3)$$

$$x_r^L \le x_r \le x_r^U; \qquad r = 1,2,\ldots,R \qquad (2.4)$$

where $W(\mathbf{x})$ is the objective function represented by the weight of the structure; $\mathbf{x}$ is the vector of design variables; ρ_m is the mass density of the mth member; l_m is the length of the mth member; x_m is the cross-sectional area of the mth member; and M is the total number of members; quantities $d_i^L, d_i^U, \sigma_m^L, \sigma_m^U$, and x_r^L are lower and upper bounds of the nodal displacements, member stresses, and design variables, respectively; constants I, G, and R are number of constrained displacement degrees of freedom, number of loading cases, and number of design variables, respectively; and functions $d_{ig}(\mathbf{x})$ and $\sigma_{mg}(\mathbf{x})$ are the ith displacement and the stress in the mth member due to the gth loading case.

Considering the displacement constraints, the gradient (sensitivity) of the ith displacement degree of freedom due to the gth loading case with respect to the mth design variable can be written as (Khot et al., 1978):

$$d_{ig,m} = -\frac{\mathbf{d}_{mg}^T \mathbf{K}_m \mathbf{y}_{im}}{x_m} \tag{2.5}$$

where $\mathbf{d}_{mg}^T$ is the transpose of the displacement vector of the mth element due to the gth loading case, $\mathbf{K}_m$ is the stiffness matrix of the mth element and $\mathbf{y}_{im}$ is the displacement vector of the mth element due to a unit load at the ith degree of freedom.

If the absolute values (magnitudes) of the lower and upper bounds on displacements are equal , we can write

$$d_i^U = -d_i^L = \overline{d}_i. \tag{2.6}$$

In this case, the displacement constraint, Eq. (2.2), is written as:

$$\left| d_{ig} \right| \le \overline{d}_i. \tag{2.7}$$

The Lagrangian function for the optimization problem can be written as:

$$L(\mathbf{x},\lambda) = W(\mathbf{x}) + \sum_{i=1}^{I}\sum_{g=1}^{G} \lambda_{ig}\left[d_{ig}(\mathbf{x}) - \overline{d}_i \right] \tag{2.8}$$

where λ_{ig} is the nonnegative Lagrange multiplier associated with the ith constrained displacement degree of freedom and the gth loading case. The Kuhn-Tucker conditions for a local optimum are (Khan et al., 1979)

$$\frac{\partial L(\mathbf{x}, \lambda)}{\partial x_m} = \frac{\partial W(\mathbf{x})}{\partial x_m} + \sum_{i=1}^{r} \sum_{g=1}^{G} \lambda_{ig} \frac{\partial d_{ig}(\mathbf{x})}{\partial x_m}; \quad m = 1, 2, \ldots, M \quad (2.9)$$

where,

$$\lambda_{ig}\left(d_{ig}(\mathbf{x}) - \bar{d}_i\right) = 0; \quad i = 1, 2, \ldots, I; \quad g = 1, 2, \ldots, G; \quad \lambda_{ig} \geq 0 \quad (2.10)$$

Using Eq. (2.5) in Eq. (2.9), solving for Lagrange multipliers, and subsequently back substituting them in Eq. (2.9), we obtain the optimality criterion for the ith nodal displacement constraint

$$1 = \left(\frac{W}{\bar{d}_i}\right)\left(\frac{\mathbf{d}_{mg}^T \mathbf{K}_m \mathbf{y}_{im}}{\rho_m l_m x_m}\right); \quad m = 1, 2, \ldots, M \quad (2.11)$$

For more than one active displacement, the following recurrence formula is used to update the design variables (Khot et al., 1978):

$$(x_m)_{j+1} = (x_m)_j \left\{ \left[\sum_{i=1}^{N_a} \left(\lambda_{ig} \frac{\mathbf{d}_{ij}^T \mathbf{K}_m \mathbf{y}_{im}}{\rho_m l_m x_m} \right) \right]^{1/\mu} \right\} \quad (2.12)$$

where j is the iteration number, N_a is the number of active displacement constraints, and μ is a relaxation factor that ensures stability for the solution. The value $1/\mu$ ranges from 0.001 to 0.2 (Khan et al.,1979). In this work, we start with $1/\mu = 0.1$; whenever the value of the objective function, W, oscillates, the value of $1/\mu$ is halved.

If the stress constraint is considered, the following recurrence formula is applied for updating the design variables:

$$\left(x_m\right)_{j+1} = \left(x_m\right)_j \left\{\max\left[\left(\frac{\sigma_{mg}}{\sigma_m^L}\right) \text{or} \left(\frac{\sigma_{mg}}{\sigma_m^U}\right)\right]\right\}; \quad m = 1,2,\ldots,M \quad (2.13)$$

If the ratio of a member stress to its limiting stress is larger than one then, Eq. (2.13) is used to redesign that member. Otherwise, Eq. (2.12) is used. A member is redesigned using only one equation at a time (Eq. 2.12 or Eq. 2.13).

2.7 STORAGE SCHEMES FOR SOLUTION OF THE SIMULTANEOUS LINEAR EQUATIONS

Two different storage schemes are investigated for the solution of the simultaneous linear equations resulting from the finite element structural analysis using the Gauss elimination method: the banded storage approach and the skyline storage scheme.

2.7.1 Banded Storage Approach

In the banded storage approach, the upper half of the stiffness matrix is stored in a rectangular matrix with a bandwidth equal to one half of the maximum difference of the degrees of freedom at member ends plus one. Careful numbering is required to achieve minimum bandwidth. This is a nontrivial task for large irregular structures. A disadvantage of the banded storage approach is that some zero values inside the band width are stored and operated in the analysis while they have no effect on the solution.

2.7.2   Skyline Storage Approach

In this method, only the stiffness matrix elements that lie below the skyline are stored in a one-dimensional array.   A minimum storage is used in this approach.   This reduces the number of operations performed on the matrix elements.   Moreover, the one-dimensional skyline storage approach lends itself to vectorization effectively.

2.8   MULTITASKING ALGORITHMS

In the finite element structural analysis, most of the time is spent in setting up and assembling the element stiffness matrices into the structure stiffness matrix, the assembling of the structure load vector, and solving the system of linear equations for displacement degrees of freedom. Thus, we will concentrate on the functions *"assemble"* that setups and assembles the structure stiffness matrix, *"load_vector"* that assembles the structure load vector, and *"solve"* that solves the resulting linear equations. Parallel algorithms are developed and compared using autotasking, microtasking, and macrotasking with the objective of improving the performance of the functions.

In this chapter, three different multitasking algorithms are presented and compared for optimization of space axial-force structures. The first algorithm, called algorithm A, is based on the use of autotasking only (see Section 2.4.3). In the second algorithm, B, microtasking directives are introduced in the loop level. No load balancing is used in this case. In the third algorithm, C, both microtasking and macrotasking are used with load balancing (Table 2.1).

2.9 EXAMPLES

We solve three space truss structures using algorithms A, B, and C. These examples are rough models of space station structures.

Example 1 is a 526-member space truss (Figure 2.1). It consists of 32 equal-span panels in the longitudinal direction and one square panel in the transverse direction. It has two simple supports at each end and two other supports at the middle of the span. Thus, it is a symmetric continuous two-span truss. The upper nodes at the middle of each span are loaded in the vertical y-direction by a 60-kip (267 kN) downward load and in x and z directions by 20-kip (89 kN) loads. The displacements of the nodes at the middle of each span in the vertical y-direction are restricted to 1/200th of the span.

Example 2 is a 1046-member space truss. The geometry of this example is the same as that of Example 1, but it has 64 panels in the longitudinal direction (twice as many as Example 1) and four supports at quarter points. The upper nodes at the middle of each span are loaded by a 60-kip (267 kN) downward load in the vertical y-direction and by 20-kip (89 kN) loads in the x and z directions. The displacements of the nodes at the middle of each span in the vertical y-direction are restricted to 1/200 times the span length.

Example 3 is a 3126-member space truss structure (Figure 2.2). The U-shape structure has three wings, each consisting of 64 equal-span panels in the longitudinal direction (similar to Example 2). It has 30 simple supports as indicated in Figure 2.2. The loading and displacement constraints are similar to those of Example 2.

Table 2.1: Multitasking algorithms for optimization of space axial-force structures with vectorization.

Algorithm A: autotasking and vectorization.
Algorithm B: microtasking and vectorization.
Algorithm C: macrotasking, microtasking, and vectorization.

1. Set the number of processors (n_p).
2. Read in the input data and the starting design variables.
3. Set $1/\mu = 0.1$, iteration = 1 and operation = 1, where operation is a factor to indicate whether this step is in the analysis stage (operation=1) or in the redesign stage (operation=2).
4. Assemble the structure stiffness matrix.
   Do concurrently:
   i - Calculate element stiffness matrices.
   A (**autotasking**); B (**microtasking**); C (**macrotasking with load balancing**)
   ii- Assemble element stiffness matrices into the structure stiffness matrix.
   A (**autotasking**); B (**microtasking with guarded regions**); C (**microtasking with guarded regions**)
5. Assemble total load vector.
   Do concurrently:
   Assemble the nodal forces into the total load vector .
   A (**autotasking and vectorization**); B (**microtasking and vectorization**); C (**microtasking with load balancing and vectorization**).
6. Apply boundary conditions.

Table 2.1 - continued

Do concurrently:

i - Update the structure stiffness matrix only if operation =1.

A (**autotasking**); B (**microtasking with guarded regions**);
C (**microtasking with guarded regions**).

ii- Update total load vector.

A (**autotasking and vectorization**); B (**microtasking and vectorization**); C (**microtasking with load balancing and vectorization**).

7. Solve the linear equations.

Do concurrently:

i - Reduce the structure stiffness matrix only if operation =1.

A (**vectorization**); B (**vectorization**); C (**vectorization**).

ii- Forward substitution.

A (**autotasking and vectorization**); B (**microtasking and vectorization**); C (**microtasking with load balancing and vectorization**).

iii- Backward substitution.

A (**autotasking and vectorization**); B (**microtasking and vectorization**); C (**microtasking with load balancing and vectorization**).

8. If operation = 1, calculate the member forces and stresses.

A (**autotasking**); B (**microtasking**); C (**microtasking with load balancing**).

If operation = 2, go to step 15.

9. Calculate the objective function (W) using Eq. (2.1).

A (**autotasking**); B (**microtasking**); C (**microtasking with load balancing**).

If the difference between the new and old objective functions is

Table 2.1 - continued

Less than 0.1 %, stop and print the results.
Otherwise, go to step 10.

10. Set operation = 2.

    If there is no constrained displacements, set maximum
    displacement ratio = 1 and go to step 11. Otherwise,
    calculate the maximum displacement ratio and go to step 11.

    A (**autotasking and vectorization**); B (**microtasking with
    guarded regions and vectorization**); C (**microtasking with
    guarded regions and vectorization**)

11. Calculate the maximum stress ratio (stress_ratio).

    A (**autotasking and vectorization**); B (**microtasking with
    guarded regions and vectorization**); C (**microtasking with
    guarded regions and vectorization**)

    If there is no constrained displacement, go to step 16.
    Otherwise go to step 12.

12. If iteration = 1, go to step 14.

    If the value of the objective function (W) is less than that of the
    previous iteration, divide the value of $1 / \mu$ by two and go to
    step14. Otherwise continue.

13. Find the active displacements, that is, those within 0.1% of the
    allowable values.

    A (**autotasking**); B (**microtasking**); C (**microtasking with
    load balancing**)

14. Apply unit loads in the directions of the most violated degrees
    of freedom one at a time; each time go to step 6.

15. Calculate the displacement gradients using Eq. (2.5)
    Do concurrently:

Table 2.1 - continued

i- Calculate the element stiffness matrices.

A (**autotasking**); B (**microtasking**); C (**microtasking with load balancing**).

ii- Calculate the displacement gradients using Eq. (2.5).

A (**autotasking**); B (**microtasking**); C (**microtasking with load balancing**).

16. Calculate the new design variables for the next iteration as follows:

    If stress_ratio > 1, modify the design variables using Eq. (2.13). Otherwise, modify the design variables using Eq. (2.12).

    A (**autotasking**); B (**microtasking**); C (**microtasking with Load balancing**).

17. Calculate the new objective function and set iteration = iteration+1.

18. Go to step 4.

2.10 SPEEDUP RESULTS

We study the speedup due to vectorization and multitasking for the three examples described in the previous section using the algorithms A, B, and C. The theoretical speedup due to multitasking is defined as the ratio of the execution time spent by a task in a sequential program to that spent in a concurrent program (CRAY, 1987).

Speedup results due to vectorization and various types of multitasking are presented for three main functions of the algorithms: *"assemble"*, *"load-vector"*, and *"solve"*. The last function is divided into two parts: reduction of the stiffness matrix into the product of a lower triangular and an upper triangular matrix, and forward and backward substitutions considered together. At the end of the chapter we present speedup results for complete solution of the optimization problem.

The relative times used in the solution of linear simultaneous equations for displacement degrees of freedom using the banded and skyline storage schemes are shown in Figure 2.3. In the figure, the time spent in the solution of the linear simultaneous equations in Example 1 (378 degrees of freedom) using the skyline approach and with the application of vectorization is chosen as a unit. The time spent in each of the other cases is represented relative to that unit time. For instance, the relative time for the solution of the linear simultaneous equations for example 1 (378 degrees of freedom) using the banded storage approach and applying vectorization is 2.2 units.

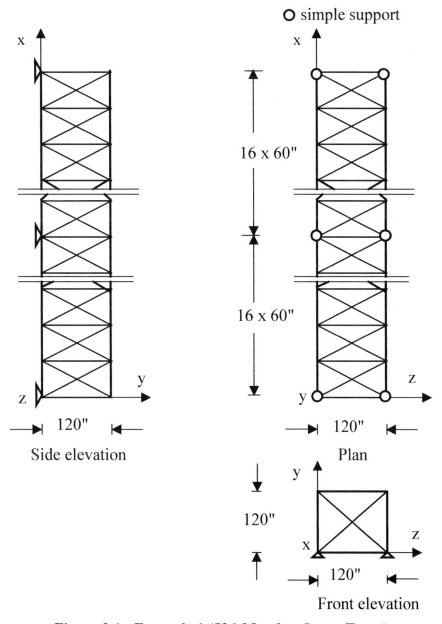

Figure 2.1: Example 1 (526-Member Space Truss)
(1 in. = 25.4 mm)

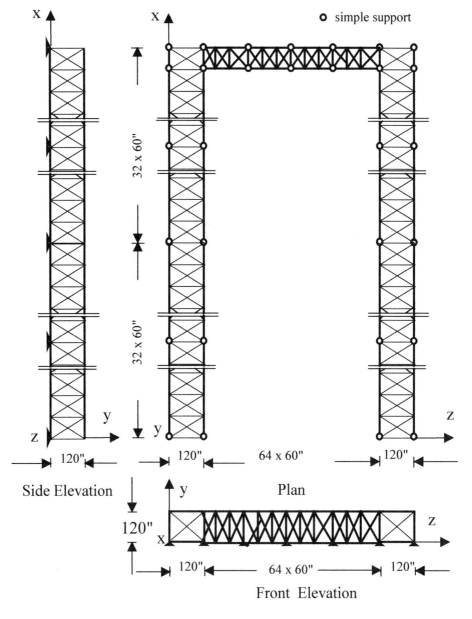

Figure 2.2: Example 3 (3126-Member Space Truss)

Figure 2.3 clearly shows that the skyline approach provides better speedups especially when vectorization is employed. The improvement in the performance is quite substantial for large structures. Thus, in the remaining part of this and subsequent chapters the skyline approach is used.

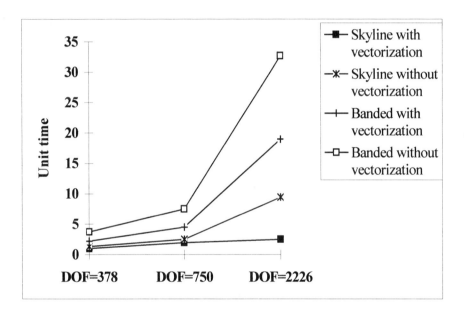

Figure 2.3: Relative times used in solution of linear simultaneous equations using banded and skyline schemes for examples 1 (DOF=378), 2 (DOF=750), and 3 (DOF= 2226)

Figures 2.4 to 2.7 show the speedup results for setting up the load vector, setting up the structure stiffness matrix, decomposing the structure stiffness matrix into the product of lower and upper triangular matrices, and forward/backward substitutions for Example 1, respectively.

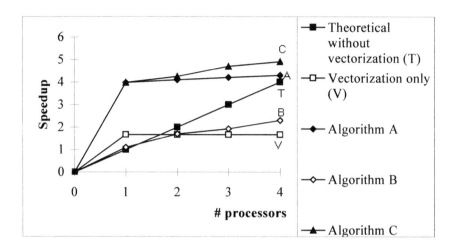

Figure 2.4: Speedup comparisons for function *"load-vector"* for Example 1

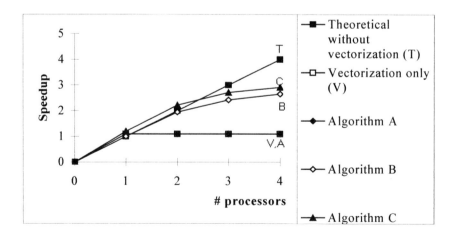

Figure 2.5: Speedup comparisons for function *"assemble"* for Example 1

Figure 2.4 shows that autotasking (algorithm A) improves the speedup for the function "*load-vector*" substantially because this function consists of only nested loops without any dependencies among matrix elements and function calls within loops. Microtasking (algorithm B) does not improve the performance of this function substantially. This is due to the relatively large overhead needed in microtasking (compared with the amount of the work done) and poor load balancing which in turn deteriorates the speedup due to vectorization. In algorithm C where microtasking is combined with macrotasking, uniform load balancing is achieved. Thus, the effect of stripmining is maximized in algorithm C.

Figure 2.5 shows that vectorization and autotasking (algorithm A) do not improve the speedup in the function "*assemble*" because of the existence of function calls and the guarded regions. Microtasking (Algorithm B) improves the speedup in the function "*assemble*" because the amount of work done in setting up the element stiffness matrices and assembling them into the structure stiffness matrix is relatively large compared with the overhead required in microtasking. Algorithm C produces higher speedup than both algorithms A and B .

Figure 2.6 shows the speedup results for decomposing the structure stiffness matrix into the product of lower and upper triangular matrices for Example 1 due to vectorization only. In this part, the steps of the solution are interdependent. As a result, vectorization has been used only for some independent short loops. Multitasking has not been used in this part because errors might occur due to the interdependency of the elements of the matrices.

Figure 2.7 shows the speedup for the second part of the function "*solve*" that performs the forward and backward substitutions

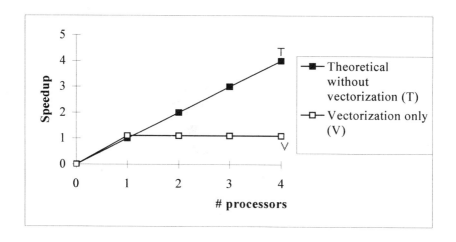

Figure 2.6: Speedup comparisons for stiffness matrix decomposition in function *"solve"* for Example 1

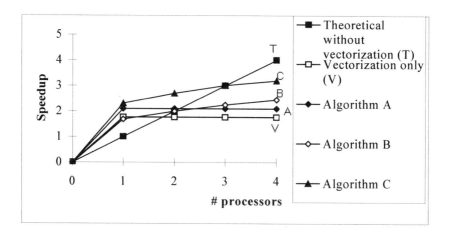

Figure 2.7: Speedup comparisons for forward and back Substitutions in function *"solve"* for Example 1

for computing the nodal displacement vector. Both vectorization and multitasking are used in this part. Algorithm C yields substantially higher speedups compared with the algorithms A and B.

Figures 2.8 to 2.11 show the speedup results for setting up the load vector, setting up the structure stiffness matrix, decomposing the structure stiffness matrix into the product of lower and upper triangular matrices, and forward/backward substitutions for Example 3. Comparing Figures 2.8 to 2.11 with the corresponding Figures 2.4 to 2.7 for Example 1, we observe that the speedups due to both vectorization and multitasking increase with the size of the structure. The increase is specially substantial for multitasking. These figures also demonstrate the superiority of algorithm C where we have employed a judicious combination of vectorization, microtasking, and macrotasking with load balancing.

In order to observe the effectiveness of the parallel processing itself without vectorization, the speedups achieved in algorithm C with and without vectorization for the functions "*load_vector*", "*assemble*", and "*solve*", are shown in Figures 2.12 to 2.14, respectively. It can be seen that when using multitasking alone the best speedup is achieved in the function "*assemble*" (Figure 2.13). This is because the amount of work done in setting up the element stiffness matrices and assembling them into the structure stiffness matrix is relatively large compared with the overhead required in multitasking. The first part of the function "*solve*" is not included in this comparison because it uses vectorization only.

The overall speedup results for the algorithm C for the complete optimization of three space truss examples without and with vectorization are presented in Figures 2.15 and 2.16, respectively. It is seen that in both cases the speedup increases substantially with

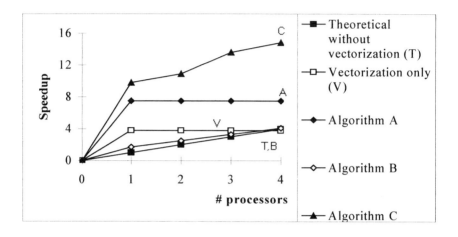

Figure 2.8: Speedup comparisons for function *"load-vector"* for Example 3

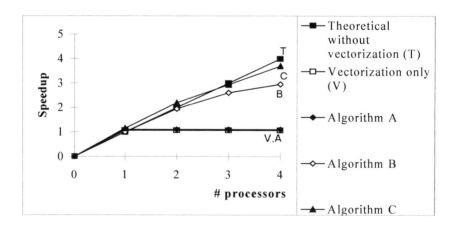

Figure 2.9: Speedup comparisons for function *"assemble"* for Example 3

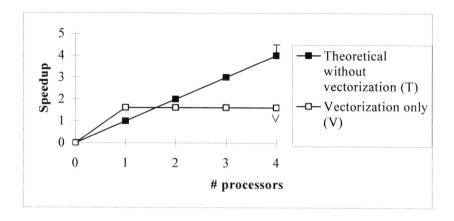

Figure 2.10: **Speedup comparisons for stiffness matrix decomposition in function "*solve*" for Example 3**

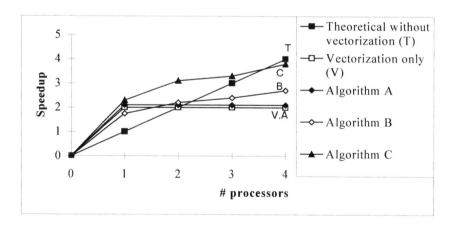

Figure 2.11: Speedup comparisons for forward and backward substitutions in function "*solve*" for Example 3

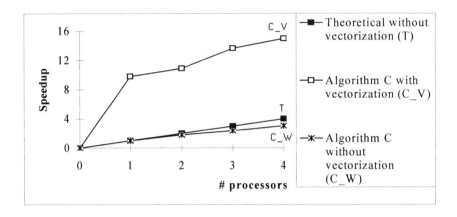

**Figure 2.12: Speedup comparisons for function "*load-vector*"
with and without vectorization for Example 3**

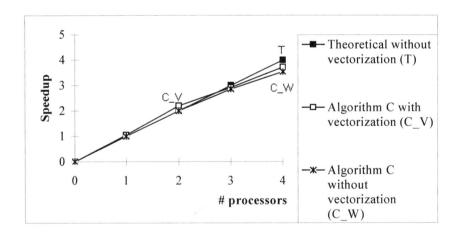

**Figure 2.13: Speedup comparisons for function "*assemble*"
with and without vectorization for Example 3**

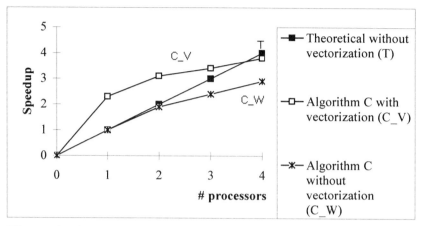

Figure 2.14: Speedup comparisons for forward and backward substitutions in function "*solve*" with and without vectorization for Example 3

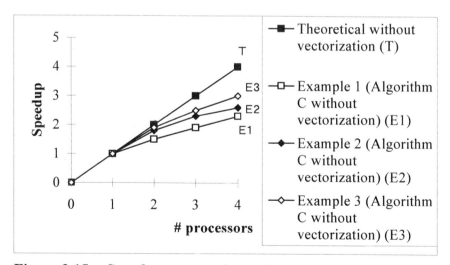

Figure 2.15: **Speedup comparisons for overall optimization using Algorithm C for three examples without vectorization**

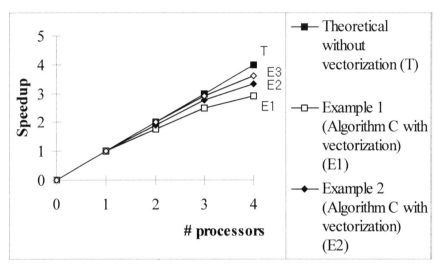

Figure 2.16: Speedup comparisons for overall optimization using Algorithm C for three examples with vectorization

the size of the structure.

2.11 CONCLUSIONS

A number of general conclusions may be drawn:

1. Autotasking works best in programs where most of the code consists of nested loops. Autotasking can not be performed when loop iterations contain interdependent array elements or when there is a function call. Basically, autotasking is effective only when the program has a simple structure.

2. Microtasking is simple to implement but does not improve the speedup substantially in complex problems without combining it with macrotasking.

3. The skyline approach is superior to the banded storage approach for storing the structure stiffness matrix in a vectorized and multitasked algorithm.

4. Judicious combination of vectorization, microtasking, and macrotasking is required· in order to develop an efficient vectorized and multitasked algorithm. Algorithm C presented in this chapter is an example of such algorithm for optimization of large structures where substantial processing power is required.

5. The processing time required for optimization of large structures increases exponentially with the size of the structure (number of design variables). Example 3 of this chapter has 3126 members and 2226 displacement degrees of freedom. Development of efficient concurrent algorithms utilizing the unique architecture and capabilities of high-performance computers results in substantial reduction in the overall execution time.

CHAPTER 3

FORMULATION OF THE INTEGRATED STRUCTURAL/CONTROL OPTIMIZATION

3.1 DYNAMIC CONTROL ANALYSIS

The discretized equations governing the controlled motion of a structure is expressed as

$$\mathbf{M\ddot{u} + C\dot{u} + Ku = Df + D_of_o} \tag{3.1}$$

where $\mathbf{M}$ is the structure mass matrix, $\mathbf{K}$ is the structure stiffness matrix, and $\mathbf{C}$ is the structure damping matrix (assumed a linear function of the mass and stiffness matrices). The matrices $\mathbf{M}$, $\mathbf{C}$, and $\mathbf{K}$ are $N \times N$ square matrices, where N is the number of displacement degrees of freedom of the structure. The quantity $\mathbf{D}$ is the $N \times N_f$ applied load distribution matrix that relates the control input vector $\mathbf{f}$ to the structure coordinates system. The number of elements in $\mathbf{f}$ is equal to the number of actuators, N_f (devices creating the control forces). The column matrix $\mathbf{u}$ is the vector of structure displacements due to dynamic loadings. The quantity $\mathbf{D_o}$ is the $N \times N_{f_o}$ applied load distribution matrix that relates the applied dynamic forces $\mathbf{f_o}$ to the structure coordinates system, where N_{f_o} is the number of applied dynamic loadings at the displacement degrees of freedom. In order to uncouple Eq. (3.1), we use the following normal mode transformation:

$$\mathbf{u} = \Phi\Psi \tag{3.2}$$

where Φ is the $N \times N$ matrix of modal shapes normalized with respect to the mass matrix and Ψ is the vector of normal mode coordinates. The columns of matrix Φ are the normalized eigenvectors of the homogeneous equations that result by setting the right hand side of Eq. (3.1) to zero. Substituting Eq. (3.2) into Eq. (3.1), premultiplying the resulting equation by Φ^T, and rearranging, the equation of motion is decoupled and can be presented by

$$\overline{\mathbf{M}}\ddot{\Psi} + \overline{\mathbf{C}}\dot{\Psi} + \overline{\mathbf{K}}\Psi = \Phi^T\mathbf{Df} + \Phi^T\mathbf{D}_\circ\mathbf{f}_\circ \tag{3.3}$$

where the matrices $\overline{\mathbf{M}}$, $\overline{\mathbf{C}}$, and $\overline{\mathbf{K}}$ are square diagonal matrices given by

$$\overline{\mathbf{M}} = \begin{bmatrix} 1 & 0 & 0 & \cdots & 0 \\ 0 & 1 & 0 & \cdots & 0 \\ \cdots & \cdots & \cdots & \cdots & \cdots \\ 0 & 0 & 0 & \cdots & 1 \end{bmatrix} \tag{3.4}$$

$$\overline{\mathbf{K}} = \begin{bmatrix} \omega_1^2 & 0 & 0 & \cdots & 0 \\ 0 & \omega_2^2 & 0 & \cdots & 0 \\ \cdots & \cdots & \cdots & \cdots & \cdots \\ 0 & 0 & 0 & \cdots & \omega_N^2 \end{bmatrix} \tag{3.5}$$

$$\overline{\mathbf{C}} = \begin{bmatrix} 2\xi_1\omega_1 & 0 & 0 & \cdots & 0 \\ 0 & 2\xi_2\omega_2 & 0 & \cdots & 0 \\ \cdots & \cdots & \cdots & \cdots & \cdots \\ 0 & 0 & 0 & \cdots & 2\xi_N\omega_N \end{bmatrix} \tag{3.6}$$

and ω_i and ξ_i are the natural frequency and damping factor for the ith mode of vibrations.

The second order uncoupled Eq. (3.3) can be reduced to a first order equation by the transformation

$$\mathbf{X} = \begin{bmatrix} \Psi \\ \dot{\Psi} \end{bmatrix} \tag{3.7}$$

where $\mathbf{X}$ is the $2N \times 1$ vector of state variables. Using the transformation Eq. (3.7), Eq. (3.3) can be written as the following first order equation:

$$\dot{\mathbf{X}} = \mathbf{A}\mathbf{X} + \mathbf{B}\mathbf{f} + \mathbf{B}_\circ \mathbf{f}_\circ \tag{3.8}$$

where $\mathbf{A}$ is a $2N \times 2N$ matrix called the plant matrix, which describes the original uncontrolled system, and $\mathbf{B}$ is the $2N \times N_f$ input matrix. The plant matrix $\mathbf{A}$ can be expressed as:

$$\mathbf{A} = \begin{bmatrix} 0 & \mathbf{M} \\ \hline -\overline{\mathbf{K}} & -\overline{\mathbf{C}} \end{bmatrix} \tag{3.9}$$

and the input matrix $\mathbf{B}$ is given by

$$\mathbf{B} = \begin{bmatrix} 0 \\ \hline \Phi^T \mathbf{D} \end{bmatrix} \tag{3.10}$$

The $2N \times N_{f_\circ}$ matrix $\mathbf{B}_\circ$ is given by

$$\mathbf{B}_\circ = \begin{bmatrix} 0 \\ \hline \Phi^T \mathbf{D}_\circ \end{bmatrix} \tag{3.11}$$

We use the Linear Quadratic Regulator (LQR) to design controllers and to obtain an optimum gain matrix $\mathbf{G}$ by minimizing a performance index J. We use the following performance index (D'Azzo, et al., 1989):

$$J = \int_0^\infty \left(\mathbf{X}^T \mathbf{Q} \mathbf{X} + \mathbf{f}^T \mathbf{R} \mathbf{f} \right) dt \tag{3.12}$$

Entities $\mathbf{Q}$ (a positive semidefinite matrix) and $\mathbf{R}$ (a positive definite matrix) are the state and control weighting matrices, respectively. Assuming equal weights, $\mathbf{Q}$ and $\mathbf{R}$ are chosen to be equal to some constant times the identity matrix. Minimizing the performance index factor J, a minimum error and energy in the design of controllers is achieved and the state feedback control law results in the following equation:

$$\mathbf{f} = -\mathbf{G} \mathbf{X} \tag{3.13}$$

where the optimum gain matrix $\mathbf{G}$ is given by

$$\mathbf{G} = \mathbf{R}^{-1} \mathbf{B}^T \mathbf{P} \tag{3.14}$$

and $\mathbf{P}$ is a positive definite matrix called the Riccati matrix. It is obtained by solving the following Riccati equation (D'Azzo et al., 1989)

$$\mathbf{Q} + \mathbf{P} \mathbf{A} + \mathbf{A}^T \mathbf{P} - \mathbf{P} \mathbf{B} \mathbf{R}^{-1} \mathbf{B}^T \mathbf{P} = 0 \tag{3.15}$$

The equations of motion without the controllers is called the open loop system of equations which is given by

$$\dot{\mathbf{X}} = \mathbf{A}\mathbf{X} + \mathbf{B}_\circ\mathbf{f}_\circ \tag{3.16}$$

The solution of Eq. (3.16) yields the open loop transient response. In order to derive the equations of motions for the closed-loop system, which includes the control system, we substitute Eq. (3.13) into Eq. (3.8) and obtain

$$\dot{\mathbf{X}} = (\mathbf{A} - \mathbf{BG})\mathbf{X} + \mathbf{B}_\circ\mathbf{f}_\circ \tag{3.17}$$

If we let

$$\overline{\mathbf{A}} = \mathbf{A} - \mathbf{BG} \tag{3.18}$$

Eq. (3.17) can be rewritten as

$$\dot{\mathbf{X}} = \overline{\mathbf{A}}\mathbf{X} + \mathbf{B}_\circ\mathbf{f}_\circ \tag{3.19}$$

Eq. (3.19) defines the closed-loop system where $\overline{\mathbf{A}}$ is the non-symmetric closed-loop matrix with a set of complex eigenvalues given by

$$\lambda_i = \hat{\sigma}_i + j\hat{\omega}_i \tag{3.20}$$

where $\hat{\sigma}_i$ and $\hat{\omega}_i$ are the real and imaginary parts of the complex closed-loop eigenvalues and $j = \sqrt{-1}$. Quantity $\hat{\sigma}_i$ is always negative in order to have an asymptotically stable system.

The damping factors associated with the closed-loop eigenvalues are defined by (Khot, 1994)

$$\hat{\xi}_i = -\frac{\hat{\sigma}_i}{\left(\hat{\sigma}_i^2 + \hat{\omega}_i^2\right)^{1/2}} \tag{3.21}$$

The solution of Eq. (3.19) yields the closed-loop transient response. After solving for the vector $\mathbf{X}$, using Eq. (3.7), the vectors Ψ and $\dot{\Psi}$ can be obtained. Then, using Eq. (3.2) the displacement vector $\mathbf{u}$ is obtained.

3.2 FORMULATION OF THE INTEGRATED STRUCTURAL/ CONTROL OPTIMIZATION

3.2.1 General

In this section, we formulate the integrated structural/control optimization problem using the optimality criteria approach. We consider constraints on stresses, displacements, and closed-loop eigenvalues and the corresponding damping factors as follows:

Minimize :

$$W(\mathbf{x}) = \sum_{m=1}^{M} \rho_m l_m x_m \tag{3.22}$$

subjected to

$$d_i^L \le r_{ig}(\mathbf{x}) + u_i(\mathbf{x},t) \le d_i^U; \quad i = 1,2,\ldots,I; \quad g = 1,2,\ldots,G \tag{3.23}$$

$$\sigma_m^L \le \sigma_{mg}(\mathbf{x}) + \sigma_m(\mathbf{x},t) \le \sigma_m^U; \quad m = 1,2,\ldots,M; \quad g = 1,2,\ldots,G \ (3.24)$$

$$x_r^L \le x_r \le x_r^U; \quad r = 1,2,\ldots,R \tag{3.25}$$

$$\hat{\omega}_j \ge \overline{\omega}_j; \quad j = 1,2,\ldots,J \tag{3.26}$$

$$\hat{\xi}_j \ge \overline{\xi}_j; \quad j = 1,2,\ldots,J \tag{3.27}$$

where $W(\mathbf{x})$ is the objective function represented by the weight of the structure, $\mathbf{x}$ is the vector of design variables, ρ_m is the mass density of the mth member, l_m is the length of the mth member, x_m is the cross-sectional area of the mth member and M is the total number of members. The quantities $d_i^L, d_i^U, \sigma_m^L, \sigma_m^U, x_r^L$, and x_r^U are the lower and upper bounds on the nodal displacements, member stresses, and design variables, respectively. The constants I and G represent the number of constrained displacements and the number of static cases of loadings, respectively. The constants $\overline{\omega}_j$ and $\overline{\xi}_j$ represent lower bounds for the imaginary part of the jth closed-loop eigenvalue and the corresponding damping factor, respectively. The functions $r_{ig}(\mathbf{x})$ and $\sigma_{mg}(\mathbf{x})$ represent the ith displacement and the stress in the mth member due to the gth loading case. The functions $\mathbf{u}_i(\mathbf{x},t)$ and $\sigma_m(\mathbf{x},t)$ are the ith structural response and the stress in the mth member due to time-dependent dynamic loading. The quantities $\hat{\omega}_j$ and $\hat{\xi}_j$ are functions for the imaginary part of the jth closed-loop eigenvalue and the corresponding damping factor. The constant J is the number of closed-loop eigenvalues.

3.2.2   Sensitivities

The derivatives of a response quantity with respect to the design variables are defined as the sensitivities of such quantity with respect to design variables. The sensitivities are very important for the optimal structural and control design of a structure. We denote the sensitivity of a variable V with respect to the design variable x_m by $V_{,m}$ where

$$V_{,m} = \frac{\partial V}{\partial x_m} \tag{3.28}$$

3.2.2.1 *Nodal Displacement Sensitivities*

The sensitivity of the displacement of the ith degree of freedom due to the gth loading condition with respect to the mth design variable can be written as (Khot et al., 1978):

$$d_{ig,m} = -\frac{\mathbf{d}_{mg}^T \mathbf{K}_m \mathbf{y}_{im}}{x_m} \qquad (repeated) \tag{2.5}$$

where $\mathbf{d}_{mg}^T$ is the transpose of the displacement vector of the mth member due to the gth loading case, $\mathbf{K}_m$ is the stiffness matrix of the mth member and $\mathbf{y}_{im}$ is the displacement vector of the mth member when a unit load is applied in the direction of the ith degree of freedom.

3.2.2.2 *Closed-loop Eigenvalue Sensitivities*

The sensitivity of the closed-loop eigenvalue λ_i with respect to design variable x_m can be expressed as

$$\hat{\lambda}_{i,m} = \mathbf{e}_i'^T \overline{\mathbf{A}}_{,m} \mathbf{e}_i \tag{3.29}$$

where $\overline{\mathbf{A}}$ is the closed-loop matrix given by Eq. (3.18), $\mathbf{e}_i'^T$ is a vector with $2N$ elements obtained by solving the following equation:

$$\mathbf{e}_i'^T \overline{\mathbf{A}} = \hat{\lambda}_i \mathbf{e}_i'^T \tag{3.30}$$

and $\mathbf{e}_i$ is a vector with $2N$ elements obtained by solving the following equation:

$$\overline{\mathbf{A}} \mathbf{e}_i = \hat{\lambda}_i \mathbf{e}_i \tag{3.31}$$

The quantities $\mathbf{e}_i'^T$ and $\mathbf{e}_i$ are normalized eigenvectors forming the matrices $\mathbf{e}'^T$ and $\mathbf{e}$ that diagonalize the matrix $\overline{\mathbf{A}}$. Substituting Eq. (3.14) into Eq. (3.18) we obtain

$$\overline{\mathbf{A}} = \mathbf{A} - \mathbf{B}\mathbf{R}^{-1}\mathbf{B}^T\mathbf{P} \tag{3.32}$$

Letting

$$\mathbf{Z} = \mathbf{B}\mathbf{R}^{-1}\mathbf{B}^T \tag{3.33}$$

Eq. (3.32) can be rewritten as

$$\overline{\mathbf{A}} = \mathbf{A} - \mathbf{ZP} \qquad (3.34)$$

Then, the sensitivity of the closed-loop matrix $\overline{\mathbf{A}}$ with respect to design variables can be expressed as

$$\overline{\mathbf{A}}_{,m} = \mathbf{A}_{,m} - \mathbf{Z}_{,m}\mathbf{P} - \mathbf{ZP}_{,m} \qquad (3.35)$$

where $\mathbf{A}_{,m}$, $\mathbf{Z}_{,m}$, and $\mathbf{P}_{,m}$ are the sensitivities of the matrices $\mathbf{A}$, $\mathbf{Z}$, and $\mathbf{P}$ with respect to the mth design variable respectively. Matrix $\mathbf{A}$ is a function of the structural frequencies ω_j^2 and matrix $\mathbf{Z}$ is a function of the modal shape matrix, Φ. Thus, the chain rule is applied to obtain the sensitivities of the matrices $\mathbf{A}$ and $\mathbf{Z}$.

The sensitivity of ω_j^2 is expressed as (Khot et al., 1985a&b)

$$\omega_{j,m}^2 = \frac{1}{x_m} \Phi_{jm}^T \left(\mathbf{K}_m - \omega_j^2 \mathbf{M}_m \right) \Phi_{jm} \qquad (3.36)$$

Similarly, the sensitivities of Φ_j can be expressed as (Khot et. al., 1985a&b)

$$\Phi_{j,m} = \frac{1}{x_m} \sum_{i=1}^{N} \frac{1}{\omega_i^2 - \omega_j^2} \Phi_{j,m}^T \left(K_m - \omega_i^2 \mathbf{M}_m \right) \Phi_{i,m} \Phi_i, \ i \neq j \ (3.37)$$

and

$$\Phi_{j,m} = \sum_{i=1}^{N} - \frac{1}{x_m} \Phi_{i,m}^T \mathbf{M}_m \Phi_{i,m} \Phi_i; \qquad i = j \tag{3.38}$$

In order to find the sensitivities of the matrix $\mathbf{P}$, we differentiate Eq. (3.15) with respect to the design variable x_m. Differentiating Eq. (3.15) and rearranging the terms we obtain

$$\mathbf{H} = \overline{\mathbf{A}}^T \mathbf{P}_{,m} + \mathbf{P}_{,m} \overline{\mathbf{A}} \tag{3.39}$$

where the left hand side, $\mathbf{H}$, is given by

$$\mathbf{H} = -\mathbf{A}_{,m}^T \mathbf{P} - \mathbf{P}\mathbf{A}_{,m} + \mathbf{P}\mathbf{Z}_{,m}\mathbf{P} \tag{3.40}$$

Eq. (3.39) is called the Lyapunov equation whose solution results in the sensitivity $\mathbf{P}_{,m}$. To solve Eq. (3.39) for $\mathbf{P}_{,m}$, we diagonalize the matrix $\overline{\mathbf{A}}$. Premultiplying Eq. (3.39) by matrix $\mathbf{e}^T$, then postmultiplying the resulting equation by the matrix $\mathbf{e}$, and using the orthonormality of matrices $\mathbf{e}$ and $\mathbf{e}'$, the matrix $\mathbf{P}_{,m}$ is found to be

$$\mathbf{P}_{,m} = \mathbf{e}'\mathbf{S}\mathbf{e}'^T \tag{3.41}$$

where the elements of the matrix $\mathbf{S}$ are obtained from the following equation:

$$S_{jk} = \frac{\hat{\mathbf{H}}_{jk}}{\hat{\lambda}_j + \hat{\lambda}_k} \tag{3.42}$$

in which

$$\hat{\mathbf{H}} = \mathbf{e}^T \mathbf{H} \mathbf{e} \tag{3.43}$$

Using the expressions for the sensitivities from Eqs. (3.36) - (3.38) and (3.41), the sensitivities of the complex eigenvalues λ_i with respect to design variables can be obtained from Eq. (3.29).

3.2.2.3 *Closed-loop Damping Parameter Sensitivities*

Using Eq. (3.21), the sensitivity of the closed-loop damping parameter, $\xi_{i,m}$, is expressed as

$$\xi_{i,m} = \frac{\left(\hat{\omega}_i \hat{\sigma}_i \hat{\omega}_{i,m} - \hat{\omega}_i^2 \hat{\sigma}_{i,m} \right)}{\left(\hat{\omega}_i^2 + \hat{\sigma}_i^2 \right)^{3/2}} \tag{3.44}$$

where $\hat{\sigma}_{i,m}$ and $\hat{\omega}_{i,m}$ are the sensitivities of the real and imaginary parts of the closed-loop eigenvalues respectively.

3.2.3 Constraints

3.2.3.1 *Nodal Displacement Constraints*

Consider the nodal displacement constraint, Eq. (3.23). Let $d_i^U = -d_i^L = \bar{d}_i$, then Eq. (3.23) can be rewritten as

$$d_{ig} = r_{ig} + u_i \leq \bar{d}_i \tag{3.45}$$

The Lagrangian function for the optimization problem is written as

$$L(\mathbf{x}, \lambda) = \mathbf{W}(\mathbf{x}) + \sum_{i=1}^{I} \sum_{g=1}^{G} \lambda_{ig} \left[d_{ig}(\mathbf{x}) - \bar{d}_i \right] \quad (\textit{repeated}) \tag{2.8}$$

where λ_{ig} is the Lagrange multipliers associated with the ith displacement degree of freedom and the gth loading case. For a local optimum we should have (Kuhn-Tucker conditions)

$$\frac{\partial L(\mathbf{x}, \lambda)}{\partial x_m} = \frac{\partial \mathbf{W}(\mathbf{x})}{\partial \mathbf{x}_m} + \sum_{i=1}^{r} \sum_{g=1}^{G} \lambda_{ig} \frac{\partial d_{ig}(\mathbf{x})}{\partial \mathbf{x}_m} \quad (\textit{repeated}) \tag{2.9}$$

subjected to

$$\lambda_{ig} \left(d_{ig}(\mathbf{x}) - \bar{d}_i \right) = 0; \quad i = 1,2,\ldots,I; \quad g = 1,2,\ldots,G \quad \lambda_{ig} \geq 0$$

$$(\textit{repeated}) \tag{2.10}$$

Substituting Eq. (2.5) into Eq. (2.10) and considering the kth displacement to be the only active displacement, we obtain

$$\rho_m l_m - \lambda_{ks} \frac{\mathbf{d}_{ms}^T \mathbf{K}_m \mathbf{y}_{km}}{x_m} = 0; \quad m = 1,2,\ldots,M \tag{3.46}$$

Adding up all the M equations and solving for λ_{ks}, we find

$$\lambda_{ks} = \frac{\displaystyle\sum_{m=1}^{M} \rho_m l_m x_m}{\displaystyle\sum_{m=1}^{M} \mathbf{d}_{ms}^T \mathbf{K}_m \mathbf{y}_m} = \frac{\mathbf{W}(\mathbf{x})}{d_{ks}} = \frac{\mathbf{W}}{\overline{d}_k} \qquad (3.47)$$

Substituting Eq. (3.47) into Eq. (3.46), we obtain

$$1 = \left(\frac{\mathbf{W}}{\overline{d}_i}\right)\left(\frac{\mathbf{d}_{mg}^T \mathbf{K}_m \mathbf{y}_{im}}{\rho_m l_m x_m}\right); \qquad m = 1,2,\ldots,M \quad (repeated) \qquad (2.11)$$

This is the optimality criterion satisfying the local minimum condition with respect to the nodal displacement constraint. Multiplying both sides of Eq. (2.11) by $(x_m)^\mu$ and taking the μth root we obtain a recurrence formula for the displacement constraint:

$$(x_m)_{j+1} = (x_m)_j \left\{ \left[\left(\frac{\mathbf{W}}{\overline{d}_k}\right)\left(\frac{\mathbf{d}_{mg}^T \mathbf{K}_m \mathbf{y}_{km}}{\rho_m l_m x_m}\right)\right]^{1/\mu} \right\} \qquad (3.48)$$

where $1/\mu$ ranges from .001 and .2.

When more than one displacements are active, the following recurrence formula is used to compute the Lagrangian multipliers:

$$(\lambda_{ig})_{j+1} = (\lambda_{ig})_j \left[\left(\frac{d_{ig}(\mathbf{x})}{\overline{d}_i}\right)\right]^\gamma \qquad (3.49)$$

where γ is taken as equal to 2 (Khan et al., 1979) and the following recurrence formula is used to calculate the design variables:

$$\left(x_m\right)_{j+1} = \left(x_m\right)_j \left\{ \left[\sum_{i=1}^{N_a} \left(\lambda_{ig} \frac{\mathbf{d}_{ij}^T \mathbf{K}_m \mathbf{y}_{im}}{\rho_m l_m x_m} \right) \right]^{1/\mu} \right\} \quad (repeated) \quad (2.12)$$

where N_a is the number of active displacements.

3.2.3.2 Stress Constraints

When the stress constraint is considered alone, the following recurrence formula is used for calculating the design variables:

$$\left(x_m\right)_{j+1} = \left(x_m\right)_j \left\{ \max\left[\left(\frac{\sigma_{mg}}{\sigma_m^L} \right) or \left(\frac{\sigma_{mg}}{\sigma_m^U} \right) \right] \right\}; \quad m = 1,2,\ldots, M$$

$$(repeated) \quad (2.13)$$

3.2.3.3 Closed-loop Eigenvalue Constraints

Considering the closed-loop eigenvalue constraint, Eq. (3.26), alone, the Lagrangian function for the optimization problem is expressed as

$$L(\mathbf{x}, \lambda) = W(\mathbf{x}) + \sum_{j=1}^{J} \lambda_j \left(\hat{\omega}_j - \overline{\omega}_j \right) \quad (3.50)$$

where λ_j is the Lagrange multiplier associated with the jth closed-loop eigenvalue. Following a procedure similar to that of nodal displacement constraint and considering the smallest closed-loop eigenvalue as the most critical one, we obtain a recurrence formula for the design variables

$$(x_m)_{j+1} = (x_m)_j \left[\left(\frac{W}{\sum\limits_{m=1}^{M} x_m \hat{\omega}_{1,m}} \right) \left(\frac{\hat{\omega}_{1,m}}{\rho_m l_m} \right) \right]^{1/\eta} \tag{3.51}$$

where η is a constant value.

3.2.3.4 Closed-loop Damping Parameter Constraints

Considering the closed-loop damping parameter constraint, Eq. (3.27), alone, the Lagrangian function for the optimization problem is expressed as

$$L(\mathbf{x}, \lambda) = W(\mathbf{x}) + \sum_{j=1}^{J} \lambda_j \left(\hat{\xi}_j - \overline{\xi}_j \right) \tag{3.52}$$

where λ_j is the Lagrange multiplier associated with the jth closed-loop damping parameter. Similar to the closed-loop eigenvalue constraint, we obtain the following recurrence formula for the design variables

$$
(x_m)_{j+1} = (x_m)_j \left[\left(\frac{W}{\displaystyle\sum_{m=1}^{M} x_m \hat{\xi}_{1,m}} \right) \left(\frac{\hat{\xi}_{1,m}}{\rho_m l_m} \right) \right]^{1/\varepsilon}
\tag{3.53}
$$

where ε is a constant value.

CHAPTER 4

PARALLEL ALGORITHMS FOR SOLUTION OF THE EIGENVALUE PROBLEM

4.1 INTRODUCTION

Our goal in this chapter is to develop efficient parallel algorithms for the solution of the eigenvalue problem of an unsymmetric matrix which is used repeatedly in the solution of the Riccati equation (3.15) as well as the solutions of both the open-loop system of equations, (3.16), and closed-loop system of equations, (3.19).

The $N \times N$ matrix $\mathbf{A}$ in the open-loop system of equations (3.16), the $N \times N$ matrix $\overline{\mathbf{A}}$ in the closed-loop system of equations (3.19), and the $2N \times 2N$ matrix $\mathbf{W}$ that is formed from the Riccati equation (3.15) in the following form:

$$\mathbf{W} = \begin{bmatrix} \mathbf{A}^T & \mathbf{Q} \\ \mathbf{BR}^{-1}\mathbf{B}^T & -\mathbf{A} \end{bmatrix} \tag{4.1}$$

are real unsymmetric matrices whose eigenvalues and eigenvectors are complex. Matrix $\mathbf{W}$ is needed for the solution of the Riccati matrix $\mathbf{P}$. The eigenvalue problem of these matrices must be solved repeatedly in the iterations of the integrated structural and control optimization problem.

4.2 APPROACHES FOR THE SOLUTION OF THE COMPLEX EIGENVALUE PROBLEM

A number of different methods have been proposed for finding the complex eigenvalues and eigenvectors of a real unsymmetric matrix. One of the early approaches is the LR method (Rutishauser, 1990). In this method, in each iteration s the unsymmetric matrix $\mathbf{A}^{(s)}$, where a superscript in parentheses refers to iteration number, is decomposed into the product of a lower triangular matrix $\mathbf{L}^{(s)}$ and an upper triangular matrix $\mathbf{R}^{(s)}$:

$$\mathbf{A}^{(s)} = \mathbf{L}^{(s)}\mathbf{R}^{(s)} \tag{4.2}$$

Then, in the next iteration, $s+1$, a new matrix $\mathbf{A}^{(s+1)}$ is obtained from the following transformation:

$$\mathbf{A}^{(s+1)} = \mathbf{L}^{(s)}\mathbf{A}^{(s)}\mathbf{R}^{(s)}. \tag{4.3}$$

The new matrix $\mathbf{A}^{(s+1)}$ is then decomposed into a lower triangular matrix $\mathbf{L}^{(s+1)}$ and an upper triangular matrix $\mathbf{R}^{(s+1)}$:

$$\mathbf{A}^{(s+1)} = \mathbf{L}^{(s+1)}\mathbf{R}^{(s+1)}, \tag{4.4}$$

and a new matrix $\mathbf{A}^{(s+2)}$ is obtained as follows:

$$\mathbf{A}^{(s+2)} = \mathbf{L}^{(s+1)}\mathbf{A}^{(s+1)}\mathbf{R}^{(s+1)} \tag{4.5}$$

This iterative process is continued until an upper triangular matrix, $\mathbf{A}^{(s+t)}$, is obtained, where $(s+t)$ is the final number of iterations. The diagonal elements of the resulting upper triangular matrix, $\mathbf{A}^{(s+t)}$, are the eigenvalues of $\mathbf{A}$.

   A popular method used in commercial packages is the QR method (Stewart, 1973, Golub and Van Loan, 1989 and Rutishauser, 1990). The QR method is similar to the LR method, but the unsymmetric matrix $\mathbf{A}$ is decomposed into the product of a unitary matrix $\mathbf{Q}$ and an upper triangular matrix $\mathbf{R}$. A unitary matrix $\mathbf{Q}$ is defined such that $\mathbf{Q}^H\mathbf{Q} = \mathbf{I}$, where H stands for the conjugate transpose. A modification to the QR method is the Double QR method (Stewart, 1973) which improves the accuracy in finding the imaginary parts of the eigenvalues by reducing the round-off errors. The QR method, however, is not amenable to effective parallel processing because of the interdependence of the transformations in each iteration (Mohammed and Walsh, 1986).

Another approach is the method of matrix iterations (Walsh, 1967, Franklin, 1968, Meirovich, 1980, Rutishauser, 1990, and Dongarra and Sidani, 1993). In this approach, first the most dominant eigenvalue(s) of an unsymmetric $n \times n$ matrix is (are) extracted (the largest one in the case of real eigenvalues and the largest conjugate eigenvalues in the case of complex eigenvalues in terms of absolute value). The original matrix is then reduced to a smaller matrix with eigenvalues the same as the remaining eigenvalues of the original matrix. This process is continued until all of the eigenvalues of the original matrix are found. An advantage of this method is its ability to handle identical eigenvalues.

Dongarra and Sidani (1993) presented a variation of the method of matrix iterations for an upper Hessenberg matrix A of order n (a matrix with zeros below its subdiagonal). They start by finding an initial eigenvalue, λ_o, of a small sub-matrix of A and its corresponding eigenvector, x_o. Then, they use this eigenvalue as an approximation to the dominant eigenvalue(s) of A, and its corresponding eigenvector(s). The Newton's method (Wilkinson and Reinsch, 1965) for finding the solution of polynomial is used to compute improved values for the dominant eigenvalue and its corresponding eigenvector. This procedure is repeated until convergence is achieved. Then, the original matrix is reduced to a smaller one with eigenvalues the same as the remaining eigenvalues of the original matrix.

4.3 THE METHOD USED IN THIS WORK

We present efficient parallel-vector algorithms for the solution of the complex eigenvalue problem using the general approach of matrix iterations. The problem can be attacked in two fundamentally different ways. In the first approach, operations are performed on complex variables directly. In the second approach, operations are performed on real variables until convergence, and complex variable are computed at the end. In this work we use the second approach because both real and complex eigenvalues are computed by the same procedure. We start with an initial vector of solutions and find improved solutions in successive iterations by a least square fit. After convergence the original $n \times n$ matrix is reduced to a smaller matrix. In our approach, matrix $\mathbf{A}$ does not have to be an upper Hessenberg matrix. In Dongarra and Sidani (1993), a real unsymmetric matrix $\mathbf{A}$ first has to be transformed to an upper Hessenberg matrix that requires additional computations. In contrast, in this work, we operate on the unsymmetric matrix directly thus reducing the overall computational time. Furthermore, Dongarra and Sidani (1993) solve a system of nonlinear polynomial equations in order to find improved solutions, while we use the least square fit which is computationally more efficient because only vector-matrix multiplication is involved followed by three inner products in the case of the real dominant eigenvalue, and five inner products in the case of the complex conjugate dominant eigenvalues, in order to find improved solutions in each iteration.

The eigenvectors are computed by solving the system of linear equations given by the characteristic equation $(\mathbf{A} - \lambda\mathbf{I})\,\mathbf{e} = 0$, where $\mathbf{I}$ is an $n \times n$ identity matrix, λ is the eigenvalue(s) of matrix

A, and **e** is the associated eigenvector(s). After all the eigenvalues of the matrix **A** are obtained by the method of matrix iterations, the method of inverse iterations is used to find the eigenvectors of the unsymmetric matrix (Wilkinson, 1965 and 1963). This approach guarantees fast convergence (Golub and Van Loan, 1989).

4.4 PARALLEL-VECTOR ALGORITHMS

In Chapter 2, we investigated various multitasking approaches for optimization of structures including microtasking (parallel processing at the outer loop level), macrotasking (parallel processing at function level) and autotasking (automatic distribution of tasks to multiple processors by the compiler). The parallel-vector algorithms for solution of the complex eigenvalue problem of a general unsymmetric matrix are presented in Figures 4.1a to g and Tables 4.1 to 4.6.

A bottleneck that degrades the parallel processing performance is the inner product of two vectors which is needed frequently. Two different approaches can be used as summarized in Figure 4.2. In the first approach, shown in Figure 4.2a, the inner product is done in a guarded region where only one processor at a time can access that region. The number of variables is divided among processors as evenly as possible. Processors compute their share of the dot product concurrently using microtasking. But, only one processor at a time adds its contribution to the inner product.

In the second approach, the processors compute their share of the dot product, similar to the first approach. Then, the summation operation is performed by one of the processors using vectorization, as shown in Figure 4.2b. In the first approach, vectorization is disabled inside the guarded region. Further, there is an overhead for

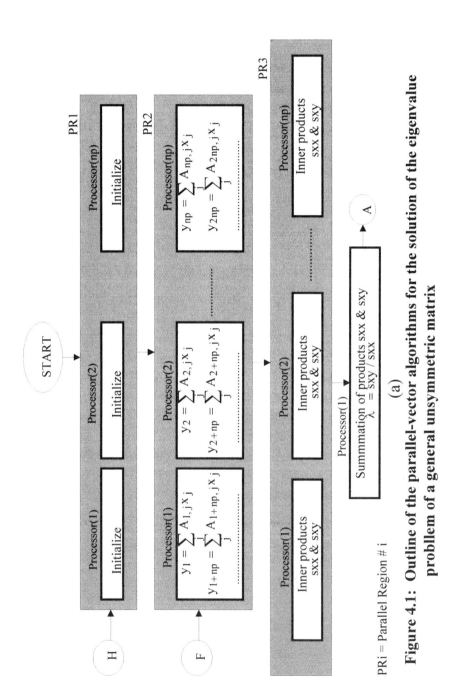

PRi = Parallel Region # i

Figure 4.1: Outline of the parallel-vector algorithms for the solution of the eigenvalue probllem of a general unsymmetric matrix

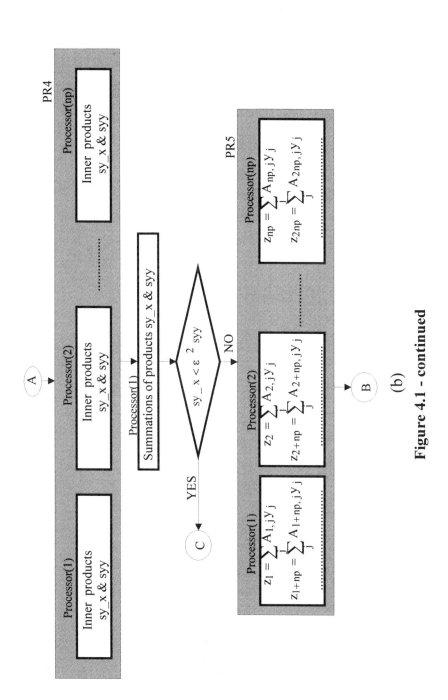

Figure 4.1 - continued

(b)

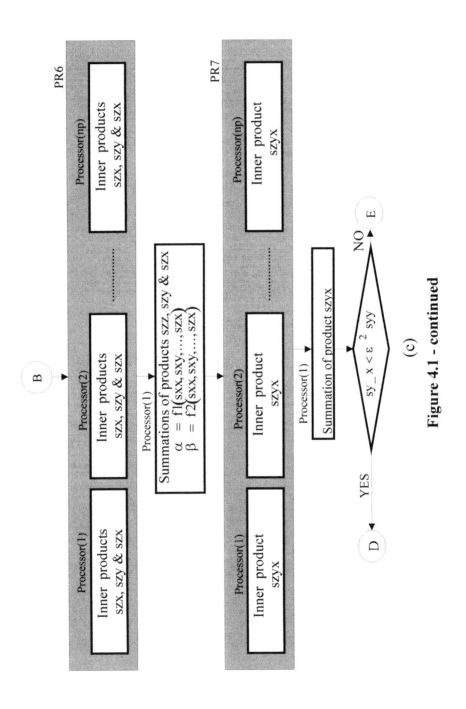

Figure 4.1 - continued

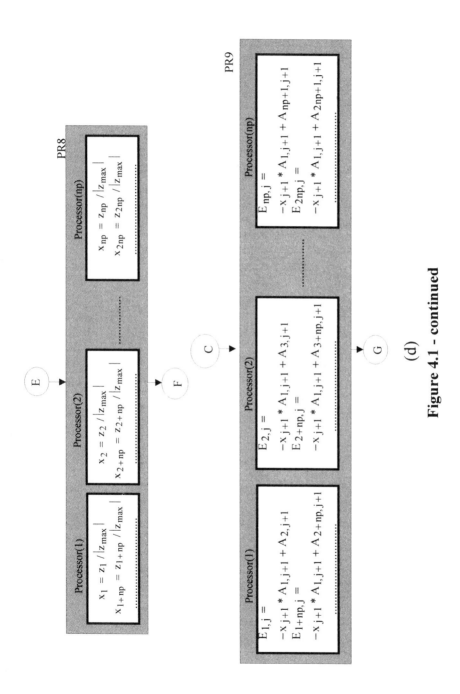

Figure 4.1 - continued

(d)

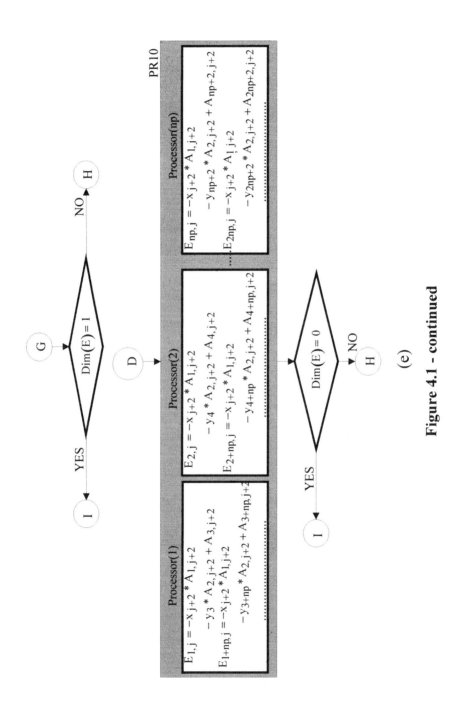

Figure 4.1 - continued

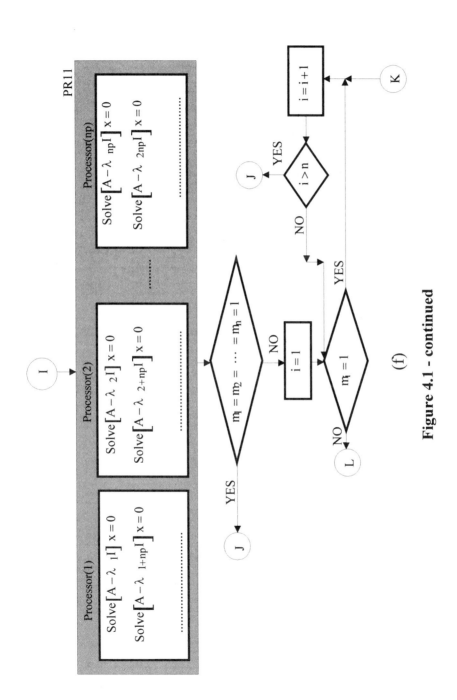

Figure 4.1 - continued

(f)

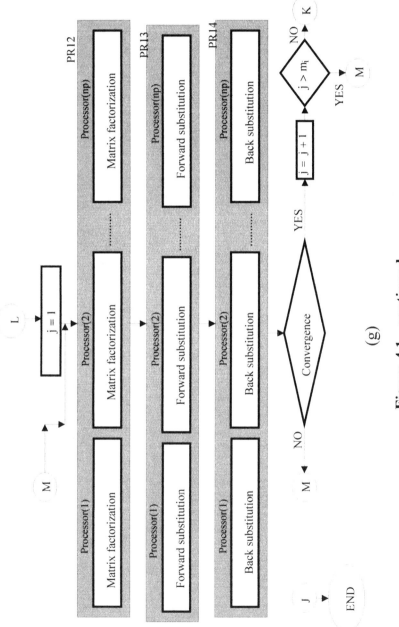

Figure 4.1 - continued

(g)

Table 4.1: Parallel algorithm to find the most dominant eigenvalue(s) and the associated eigenvector(s) using method of matrix iterations

Assume **A** is an unsymmetric $n \times n$ real matrix. We find the most dominant real eigenvalue or complex conjugate eigenvalue(s) and the associated eigenvector(s) concurrently as follows:

Set iteration k =1
1. Select an $n \times 1$ initial vector $\mathbf{x} = \mathbf{x}^1$ as follows:
     a- Divide the number of elements in **x**, n, as evenly as possible among number of processors, n_p. The number of elements in each set is num_per_set.
     b- For processor = 1 until n_p do concurrently
         (***microtasking***)
              For **element** = 1 **until** num_per_set **do**
              (***vectorization***)
                   $j =$ (processor - 1) $\times$ num_per_set + element.
                   If $j > n$, go to step 2.

                   Otherwise, $x_j^1 = x_j^0$ (initial value)   (4.6)

              Next element
         Next processor
2. Find an improved solution for eigenvector $\mathbf{y} = \mathbf{Ax}$.
     (***Matrix multiplication is done concurrently as in Table 4.4***)
3. Make a least-squares fit of **x** to **y** by choosing a real number λ to minimize

Table 4.1 – continued

$$\|\mathbf{y} - \lambda\mathbf{x}\|^2 = (\mathbf{y} - \lambda\mathbf{x})\cdot(\mathbf{y} - \lambda\mathbf{x}) = \sum_{j=1}^{n}\left(y_j - \lambda x_j\right)^2 \qquad (4.7)$$

a- Calculate the inner product **x.y** *concurrently as in Table 4.5*.

b- Calculate the inner product **x.x** *concurrently as in Table 4.5*.

c- Calculate the real number λ

$$\lambda = \frac{\mathbf{x.y}}{\mathbf{x.x}} = \frac{\displaystyle\sum_{j=1}^{n} x_j y_j}{\displaystyle\sum_{i=1}^{n} x_i^2} \qquad (4.8)$$

4. Define a small tolerance $\varepsilon_0 > 0$.
5. Check if we have a real dominant eigenvalue , λ_1 , where

$$\pm\lambda_1 = |\lambda_1| > |\lambda_2| \geq \cdots \geq |\lambda_v| \qquad (4.9)$$

a- Calculate the inner product $(\mathbf{y} - \lambda\mathbf{x})\cdot(\mathbf{y} - \lambda\mathbf{x})$ *concurrently as in Table 4.5*.

b- Calculate the inner product **y.y** *concurrently as in Table 4.5*.

c- If $(\mathbf{y} - \lambda\mathbf{x})\cdot(\mathbf{y} - \lambda\mathbf{x}) < \varepsilon_0^2 \mathbf{y}\cdot\mathbf{y}$, i.e.

Table 4.1 – continued

$$\|\mathbf{y} - \lambda\mathbf{x}\|^2 < \varepsilon_0^2 \mathbf{y} \cdot \mathbf{y} , \text{ then} \qquad (4.10)$$

we have a real eigenvalue, $\lambda_1 = \lambda$ and continue.
Otherwise go to step 6.

d- Compute the associated eigenvector $\mathbf{x}^{k+1} = \lambda_1 \mathbf{x}^k$
__concurrently as in Table 4.6__.

e- STOP.

6. Find an improved vector that is spanned by the complex eigenvectors

$\mathbf{z} = \mathbf{Ay}$. (*__Matrix multiplication is done concurrently as in__*
__Table 4.4__)

7. Check if we have complex-conjugate dominant
eigenvalues,
$\lambda_1 = \alpha + i\beta$ and $\lambda_2 = \alpha - i\beta$ where,
$|\lambda_1| = |\lambda_2| > |\lambda_3| \geq \cdots \geq |\lambda_v|$,

by finding real numbers α and β which minimize

$$\|\mathbf{z} + \alpha\mathbf{y} + \beta\mathbf{x}\|^2 = \|\mathbf{z}\|^2 + \alpha^2\|\mathbf{y}\|^2 + \beta^2\|\mathbf{x}\|^2 + 2\alpha\mathbf{z} \cdot \mathbf{y} + \qquad (4.11)$$
$$2\beta(\mathbf{z} \cdot \mathbf{x}) + 2\alpha\beta\mathbf{y} \cdot \mathbf{x}$$

as follows:

a- Compute the inner-products $\|\mathbf{x}\|^2, \|\mathbf{y}\|^2, \mathbf{y}.\mathbf{x}$, and $\mathbf{z}.\mathbf{x}$
__concurrently, as in Table 4.5__.

b- Calculate values of α and β

Table 4.1 – continued

$$\begin{bmatrix} \alpha \\ \beta \end{bmatrix} = \frac{-1}{\|\mathbf{x}\|^2 \|\mathbf{y}\|^2 - (\mathbf{y} \cdot \mathbf{x})^2} \begin{bmatrix} \|\mathbf{x}\|^2 & -\mathbf{y} \cdot \mathbf{x} \\ -\mathbf{y} \cdot \mathbf{x} & \|\mathbf{y}\|^2 \end{bmatrix} \qquad (4.12)$$

(***Matrix multiplication is done concurrently as in
Table 4.4***)
(***Inner-product is done concurrently as in Table 4.5***)

b- Compute the inner product $\|z + \alpha\mathbf{y} + \beta\mathbf{x}\|$ and $\|\mathbf{z}\|$
   concurrently as in Table 4.5.

c- If $\|z + \alpha\mathbf{y} + \beta\mathbf{x}\| < \varepsilon_0^2 \|\mathbf{z}\|^2$, then we have two complex-
   conjugate eigenvalues, λ_1 and λ_2, continue to (d).
   Otherwise, go to step 8.

d- Compute the complex-conjugate eigenvalues,

$$\lambda_1 = \frac{1}{2}\left(-\alpha + i\sqrt{4\beta - \alpha^2}\right) \quad \text{and} \quad \lambda_2 = \bar{\lambda}_1 \qquad (4.13)$$

e- Compute the associated eigenvectors, $\mathbf{x}_1$ and $\mathbf{x}_2$

$$\text{Real}(\mathbf{x}_1) = \frac{1}{2}\mathbf{x} \qquad (4.14)$$

(***Normalization is done concurrently as in Table 4.6***)

$$\text{Im}(\mathbf{x}_1) = -\frac{\alpha}{2\sqrt{4\beta - \alpha^2}}\mathbf{x} - \frac{1}{\sqrt{4\beta - \alpha^2}}\mathbf{y} \quad \text{and}$$

$$\mathbf{x}_2 = \bar{\mathbf{x}}_1 \qquad (4.15)$$

Table 4.1 – continued

(*Normalization is done concurrently as in Table 4.6*)

8. Compute a new $n \times 1$ vector $\mathbf{x} = z_m^{-1}\mathbf{z}$, where z_m is the largest absolute value in $\mathbf{z}$.

(*Normalization is done concurrently as in Table 4.6*)

Set iteration = iteration +1, go to step 2.

Table 4.2: Parallel algorithm to reduce a general unsymmetric real matrix, A, to one without the dominant eigenvalue(s).

Assume $\mathbf{A}$ is a general unsymmetric $n \times n$ real matrix. The dominant eigenvalue(s) λ_1 (real case) or λ_1 and λ_2 (conjugate-complex case) and the associated eigenvector(s) are found by the algorithm in Table 4.1 Matrix $\mathbf{R}$ is a reduced matrix whose eigenvalues are the remaining eigenvalues of matrix $\mathbf{A}$.

1. If the dominant eigenvalue of $\mathbf{A}$, λ_1, and the associated eigenvector, $\mathbf{x}$, are real, continue to step 2.
   Otherwise, go to step 7.
2. Compute new vector $\hat{\mathbf{x}}$

$$a\text{-} \quad \hat{\mathbf{x}} = \frac{1}{x_m} \mathbf{x} \qquad\qquad (4.16)$$

where $|x_m| = \max|x_i|$, $(i = 1,2,........, n)$
   (***Normalization is done concurrently as in Table 4.6***)

$$b\text{-} \quad \hat{\mathbf{x}} = \mathbf{P}_{1m} \cdot \hat{\mathbf{x}} \qquad\qquad (4.17)$$
   where

$$\mathbf{P}_{1m} = \begin{bmatrix} 0 & 0 & \cdot & 1 & 0 \\ 0 & 1 & \cdot & 0 & 0 \\ \cdots & \cdots & \cdots & \cdots & \cdots \\ 1 & 0 & \cdot & 0 & 0 \\ \cdots & \cdots & \cdots & \cdots & \cdots \\ 0 & 0 & \cdot & 0 & 1 \end{bmatrix} \qquad\qquad (4.18)$$

Table 4.2 – continued

Matrix P_{1m} permutes the first and mth column components.
(***Matrix multiplication is done concurrently as in Table 4.4***)

3. Compute new matrix $\hat{A}$

   a- $A_1 = P_{1m} A$ $\hspace{3cm}$ (4.19)

   (***Matrix multiplication is done concurrently as in Table 4.4***)

   b- $A = A_1 P_{1m}$ $\hspace{3cm}$ (4.20)

   (***Matrix multiplication is done concurrently as in Table 4.4***)

4. Form concurrently the nonsingular $n \times n$ matrix T

$$T = \begin{bmatrix} \hat{x}_1 & 0 & 0 & \cdot & 0 \\ \hat{x}_2 & 1 & 0 & \cdot & 0 \\ \hat{x}_3 & 0 & 1 & \cdot & 0 \\ \cdots & \cdots & \cdots & \cdots & \cdots \\ \hat{x}_n & 0 & 0 & \cdot & 1 \end{bmatrix} \hspace{2cm} (4.21)$$

where

$$T^{-1}\hat{A}T = \begin{bmatrix} \lambda_1 & \cdots \\ 0 & \\ 0 & R \\ \cdot & \\ 0 & \end{bmatrix} \quad (\textit{\underline{microtasking}}) \hspace{1cm} (4.22)$$

Table 4.2 – continued

The matrix $\mathbf{R}$ is the reduced matrix.

5. Compute the reduced $(n-1) \times (n-1)$ matrix $\mathbf{R}$

$$\mathbf{R} = \begin{bmatrix} -\hat{x}_2 & 1 & 0 & \cdot & 0 \\ -\hat{x}_3 & 0 & 1 & \cdot & 0 \\ \cdots & \cdots & \cdots & \cdots & \cdots \\ -\hat{x}_n & 0 & 0 & \cdot & 1 \end{bmatrix} \hat{\mathbf{A}} \begin{bmatrix} 0 & 0 & \cdot & 0 \\ 1 & 0 & \cdot & 0 \\ 0 & 1 & \cdot & 0 \\ \cdots & \cdots & \cdots & \cdots \\ 0 & 0 & \cdot & 1 \end{bmatrix} \qquad (4.23)$$

(Matrix multiplication is done concurrently as in Table 4.4)

6. Let $n = n - 1$, $\mathbf{x} = \hat{\mathbf{x}}$, and $\mathbf{A} = \mathbf{R}$.
   If $n = 1$, set $\lambda_n = R_{11}$ and STOP.
   Otherwise, go to step 11.

7. The dominant eigenvalues of $\mathbf{A}$ are complex-conjugate λ_1 and λ_2 and the associated eigenvectors are $\mathbf{x}$ and $\mathbf{y}$. Compute new vectors $\hat{\mathbf{x}}$ and $\hat{\mathbf{y}}$ as follows:

   a- Let $\mathbf{Q}_1$ be a 2×2 matrix, $\mathbf{I}$ be a 2×2 identity matrix, and $\mathbf{N}_1$ be an $n \times n$ matrix, where

$$\mathbf{Q}_1 = \frac{1}{x_m} \mathbf{I},$$

$$\mathbf{N}_1 = \mathbf{P}_{1m}, \text{ and}$$

$$|x_m| = \max|x_i| ; \ i = 1,\ldots,n \qquad (4.24)$$

Table 4.2 – continued

Compute

$$\mathbf{B} = \mathbf{N}_1[\mathbf{xy}]\mathbf{Q}_1 = \begin{bmatrix} 1 & \eta \\ * & * \\ . & . \\ * & * \end{bmatrix} \tag{4.25}$$

(_Matrix multiplication is done concurrently as in Table 4.4_)

where $[\mathbf{x} \ \mathbf{y}]$ is an $n \times 2$ matrix with the eigenvectors $\mathbf{x}$ and $\mathbf{y}$ as its columns.

b- Let $\mathbf{Q}_2$ be a 2×2 matrix defined by

$$\mathbf{Q}_2 = \begin{bmatrix} 1 & -\eta \\ 0 & 1 \end{bmatrix} \tag{4.26}$$

Compute

$$\mathbf{C} = \mathbf{BQ}_2 = \begin{bmatrix} 1 & 0 \\ * & \xi_2 \\ . & . \\ * & \xi_n \end{bmatrix} \tag{4.27}$$

(_Matrix multiplication is done concurrently as in Table 4.4_)

c- Let $|\xi_k| = \max|\xi_i|(i \geq 2)$ and the $n \times n$ matrices
$\mathbf{N}_2 = \mathbf{P}_{2k}$ and $\mathbf{N}_3 = diagonal(1, 1/\xi_k, 1, \ldots, 1) \tag{4.28}$

Table 4.2 – continued

Compute

$$D = N_3 N_2 C = \begin{bmatrix} 1 & 0 \\ \omega & 1 \\ * & * \\ . & . \\ * & * \end{bmatrix} \tag{4.29}$$

(***Matrix multiplication is done concurrently as in Table 4.4***)

d- Let Q_3 be a 2×2 matrix:

$$Q_3 = \begin{bmatrix} 1 & 0 \\ -\omega & 1 \end{bmatrix} \tag{4.30}$$

Compute

$$[\hat{x} \, \hat{y}] = DQ_3 \tag{4.31}$$

(***Matrix multiplication is done concurrently as in Table 4.4***)

8. Compute new matrix $\hat{A}$:

$$\begin{aligned}\hat{A} = \text{diagonal} \left(1, 1/\xi_k, 1, \ldots, 1 \right) \, P_{2k} P_{1m} A P_{1m} P_{2k} \\ \text{diagonal} \left(1, \xi_k, 1, \ldots, 1 \right) \end{aligned} \tag{4.32}$$

(***Matrix multiplication is done concurrently as in Table 4.4***)

9. Compute the reduced $(n - 2) \times (n - 2)$ matrix R

Table 4.2 – continued

$$
\mathbf{R} =
\begin{bmatrix}
-\hat{x}_3 & -\hat{y}_3 & 1 & \cdot & 0 \\
\cdots & \cdots & \cdots & \cdots & \cdots \\
-\hat{x}_n & -\hat{y}_n & 0 & \cdot & 1
\end{bmatrix}
\begin{bmatrix}
\hat{a}_{13} & \hat{a}_{14} & \cdot & \hat{a}_{1n} \\
\cdots & \cdots & \cdots & \cdots \\
\hat{a}_{n3} & \hat{a}_{n4} & \cdot & \hat{a}_{nn}
\end{bmatrix}
\quad (4.33)
$$

*(**Matrix multiplication is done concurrently as in Table 4.4**)*

10. Let $n = n - 2$, $\mathbf{x} = \hat{\mathbf{x}}$, $\mathbf{y} = \hat{\mathbf{y}}$, and $\mathbf{A} = \mathbf{R}$.
    If $n = 1$, set $\lambda_n = R_{11}$ and STOP. Else if $n = 0$, STOP.

    Otherwise, go to step 11.

11. Compute the most dominant eigenvalue(s) for matrix $\mathbf{A}$
    (as in Table 4.1).
    Go to step 1.

Table 4.3: Parallel algorithm for finding the eigenvectors associated with distinct and repeated eigenvalues of a general unsymmetric $n \times n$ real matrix A

Input: n, $\lambda_1 \cdots\cdots \lambda_n$, A, n_p, $m_1 \cdots\cdots m_n$, and ε

FOR $p = 1$ **UNTIL** n_p **DO**   **(Microtasking)**

  $j = p$

  (a) $\mathbf{b}_1 = \begin{bmatrix} 0 \\ \overline{\mathbf{x}}(j) \end{bmatrix}$ **(Vectorization)** , $\tau = \lambda_j$

  (b) $k = 1$

  (c) $\mathbf{b}_k = (\mathbf{A} - \tau \mathbf{I})^{-1} \mathbf{y}$ **(Macrotasking and Vectorization)**

   $\mathbf{b}_{k+1} = \dfrac{\mathbf{y}}{|\mathbf{y}|}$ **(Vectorization)**

  IF $\left| \mathbf{b}_{k+1} - \mathbf{b}_k \right| < \varepsilon$ **THEN**

   $\mathbf{e} = \mathbf{b}_{k+1}$, $j = j + n_p$

   IF $j > n$ **THEN**

     Go to (d)

   ELSE

     Go to (a)

  ELSE

     Continue

   $\mathbf{b}_k = \mathbf{b}_{k+1}$ **(Vectorization)**, $k = k + 1$

   IF $k > 5$ **THEN**

     $\tau = \tau + \dfrac{1}{\mathbf{b}_k \cdot \mathbf{y}}$ **(Vectorization)**

Table 4.3 – continued

       Go to (b)
    ELSE
       Go to (c)
NEXT p
(d) $i = 1$, $r = 2$
(e) $\mathbf{e}_1 = \mathbf{x}_i$ (**Vectorization**)
  FOR $j = 1$ **UNTIL** $r - 1$ **DO**

$$\mathbf{e}_{rj} = \mathbf{e}_r \cdot \mathbf{e}_j \quad \text{(\textbf{Microtasking and Vectorization})}$$

$$\mathbf{e}_r = \mathbf{e}_r - \frac{\mathbf{e}_{rj}}{\|\mathbf{e}_j\|}\, \mathbf{e}_j \quad \text{(\textbf{Vectorization})}$$

  NEXT j
  $\mathbf{b}_1 = \mathbf{e}_r$ (**Vectorization**), $\tau = \lambda_i$
(f) $k = 1$
(g) $\mathbf{b}_k = (\mathbf{A} - \tau\mathbf{I})^{-1}\mathbf{y}$ (**Microtasking and Vectorization**)

$$\mathbf{b}_{k+1} = \frac{\mathbf{y}}{|\mathbf{y}|} \quad \text{(\textbf{Microtasking and Vectorization})}$$

 IF $|\mathbf{b}_{k+1} - \mathbf{b}_k| < \varepsilon$ **THEN**
    $\mathbf{e} = \mathbf{b}_{k+1}$, $r = r + 1$
   IF $r > m_i$ **THEN**
      $i = i + 1$, $r = 2$
     IF $i > n$ **THEN**
       Stop
     ELSE
       Go to (e)

Table 4.3 – continued

ELSE
    Go to (e)
ELSE
    Continue
$k = k + 1$
FOR $j = 1$ **UNTIL** r -1 **DO**

$$\mathbf{b}_{kj} = \mathbf{b}_{k+1} \cdot \mathbf{e}_j \quad \textbf{(Microtasking and Vectorization)}$$

$$\mathbf{b}_k = \mathbf{b}_{k+1} - \frac{\mathbf{b}_{kj}}{\|\mathbf{e}_j\|} \mathbf{e}_j \quad \textbf{(Vectorization)}$$

NEXT j
IF $k > 5$ **THEN**

$$\tau = \tau + \frac{1}{\mathbf{b}_k \cdot \mathbf{y}} \quad \textbf{(Microtasking and Vectorization)}$$

    Go to (f)
ELSE
    Go to (g)
$i = i + 1$
STOP

Table 4.4: Parallel algorithm for multiplication of two matrices

Assume **A**, **B**, and **C** are $I \times J$, $J \times K$, and $I \times K$ matrices, respectively. The parallel algorithm for matrix multiplication $\mathbf{C} = \mathbf{AB}$ is written as follows:

1. Divide number of rows of matrix **A**, I, as evenly as possible, among the number of processors, n_p. Let the number of rows per processor be num_per_set.
2. **FOR** $p = 1$ **UNTIL** n_p **DO** concurrently   (***microtasking***)
   FOR element = 1 **UNTIL** num_per_set **DO** sequentially
   a- Set $i = ($ processor-1$) \times$ num_per_set + element
   b- **IF** $i > I$, **THEN** go to step 3. **ELSE,**
       FOR $k = 1$ **UNTIL** K **DO**
          $C_{ik} = 0$
          FOR $j = 1$ **UNTIL** J **DO** (***vectorization***)
             $C_{ik} = C_{ik} + A_{ij} \times B_{jk}$
          NEXT j
       NEXT k
   NEXT element.
   NEXT p
3. **STOP.**

Table 4.5: Parallel algorithm for inner-product of two vectors.

Suppose **A** and **B** are two column matrices with I rows. The inner product $C = \mathbf{A} \bullet \mathbf{B}$ is found concurrently as follows:

1. Divide number of rows, I, as evenly as possible, among the number of processors, n_p. Let the number of rows per processor be num_per_set.
2. $C = 0$.
3. **FOR** $p = 1$ **UNTIL** n_p **DO** concurrently   (***microtasking***)
    $C_p = 0$
    FOR element $= 1$ **UNTIL** num_per_set **DO** (***vectorization***)
        a- Set $i = ($ processor - 1 $) \times$ num_per_set + element
        b- **IF** $i > I$, then go to step 4.
           ELSE, continue.
        c- $C_p = C_p + A_i \times B_i$
    NEXT element.
    DO sequentially
    $C = C + C_p$ (***guarded region***)
    NEXT p
4. **STOP**

Table 4.6: Parallel algorithm for normalization of a vector with respect to a constant.

Suppose **A** is a column matrix with I elements. The normalization of **A** with respect to a constant K is done concurrently as follows:

1. Divide number of rows, I, as evenly as possible among number of processors, n_p. Let number of elements per processor be num_per_set.
2. **FOR** $p = 1$ **UNTIL** n_p **DO** concurrently   (***microtasking***)
     FOR element = 1 **UNTIL** num_per_set **DO** (***vectorization***)
         a- Set $i = (p - 1) \times$ num_per_set + element
         b- **IF** $i > I$, then go to step 3.
             ELSE $A_i = A_i / K$
     NEXT element.
     NEXT p
3. **STOP**

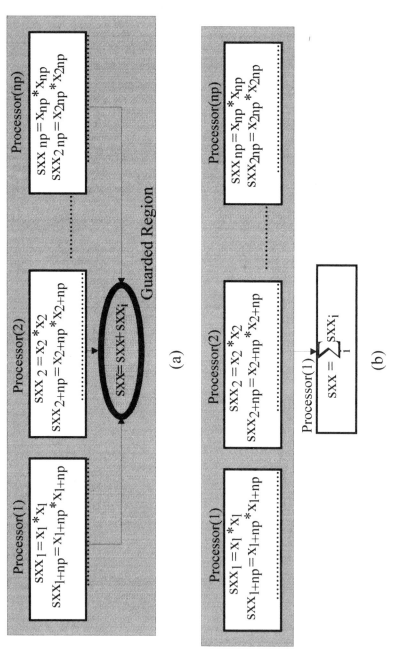

Figure 4.2: Approaches for the parallel computation of the inner product.

synchronization. We found the second approach to be substantially more efficient for large problems and is consequently used in the algorithms presented in this book.

There is also another approach to the inner product computation, that is performing the entire inner product as a vector operation on one processor while the others are waiting. We found this approach to be less efficient than the approach described in the previous paragraph.

A major portion of the computation time is consumed by the solution of systems of linear equations needed in the method of inverse iterations for finding the eigenvectors. Two different concurrent processing approaches are used in order to achieve maximum efficiency. After all the eigenvalues are obtained, they are classified into distinct and repeated eigenvalues. For the eigenvectors of m distinct eigenvalues we have m independent systems of n linear equations. For this part, we distribute the m systems of linear equations among n_p processors as evenly as possible using macrotasking (Figure 4.1f). Each system of linear equations is solved on a separate processor using vector operations.

For the remaining repeated n-m eigenvalues, the possibility of convergence to the previously found eigenvector(s) must be avoided. For these repeated eigenvalues each set of linear equations corresponding to one eigenvalue is solved concurrently using microtasking as well as vectorization (Figure 4.1g).

The parallel-vector or stripmining algorithms outlined in Figures 4.1a to g consist of 14 parallel regions to be described shortly. Between these parallel regions there exists some operations that are performed by one processor, called master processor and identified as processor (1) in Figure 4.1. These operations which are kept at a minimum for maximum parallel processing speedup include the summation in the inner product operation, convergence check, checking for eigenvalue multiplicity, etc. They are identified in Figure 4.1. A short description of each parallel region follows:

PR1: initialization of vector $\mathbf{x} = \mathbf{x}^{(k)}$ (Figure 4.1a).

PR2: computation of the first improved vector $\mathbf{y} = \mathbf{x}^{(k+1)}$ (Figure 4.1a).

PR3: computation of the inner products $sxx = \mathbf{x} \cdot \mathbf{x}$ and $sxy = \mathbf{x} \cdot \mathbf{y}$ (Figure 4.1a).

PR4: computation of the inner products $sy\_x = (y - \lambda\mathbf{x}) \cdot (y - \lambda\mathbf{x})$ and $syy = \mathbf{y} \cdot \mathbf{y}$ (Figure 4.1b).

PR5: computation of the second improved vector $\mathbf{z} = \mathbf{x}^{(k+2)}$ (Figure 4.1b).

PR6: computation of the inner products $szz = \mathbf{z} \cdot \mathbf{z}$, $szy = \mathbf{z} \cdot \mathbf{y}$, and $szx = \mathbf{z} \cdot \mathbf{x}$ (Figure 4.1c).

PR7: computation of the inner product
$szyx = (\mathbf{z} + \alpha\mathbf{y} + \beta\mathbf{x}) \cdot (\mathbf{z} + \alpha\mathbf{y} + \beta\mathbf{x})$ (Figure 4.1c).

PR8: normalization of vector $\mathbf{z}$ with respect to its length to find a new improved vector (to minimize the roundoff errors) (Figure 4.1d).

PR9: computation of the reduced $(n-1) \times (n-1)$ matrix in the case of real dominant eigenvalue (Figure 4.1d).

PR10: computation of the reduced $(n-2) \times (n-2)$ matrix in the case of conjugate complex dominant eigenvalues (Figure 4.1e).

PR11: computation of the eigenvectors associated with the distinct eigenvalues (Figure 4.1f).

PR12: computation of the eigenvectors associated with the
to   repeated eigenvalues (Figure 4.1g).
PR14

4.5 EXAMPLES

Several Examples are presented in order to demonstrate the accuracy and efficiency of the parallel-vector algorithms presented in this chapter. The first two Examples are obtained from small structures solved and reported in the literature by other investigators. These examples are used as test examples to verify the results obtained from the parallel-vector algorithms presented in this book. Examples 3 to 6 are newly created with the objective of demonstrating the robustness and efficiency of the parallel-vector algorithms.

Efficiency is presented in terms of millions of floating point operations per second (MFLOPS) and the speedup. The vectorization speedup is defined as the ratio of the single-processor execution time of the vectorized algorithm to the corresponding algorithm without vectorization executed on one processor. The parallel processing speedup is defined as the ratio of the execution time of the sequential algorithm on a single processor to that of the concurrent algorithm on n_p processors.

4.5.1 Example 1

This example is a 4×4 real unsymmetric closed-loop matrix $\overline{A}$ resulting from the two-bar truss shown in Figure 4.3 and solved by Khot (1994). The truss has 4 state variables (two displacements and two velocities at node 2). One actuator and a sensor are collocated along member 1 between nodes 1 and 2. The modulus of elasticity of the members and the unit weight of the structural material are 1.0 and 0.001 in unspecified units, respectively. The cross-sectional areas of members 1 and 2 are chosen as 1000 and

100 units, respectively. A lumped mass of 2 units is attached to node 2. The weighting 4×4 **Q** and 1×1 **R** matrices are chosen as identity matrix. The closed-loop matrix is given as

$$
\overline{\mathbf{A}} = \begin{bmatrix} 0.0 & 0.0 & 1.0 & 0.0 \\ 0.0 & 0.0 & 0.0 & 1.0 \\ -1.3766000 & -0.0022523 & -0.0456820 & -0.0355300 \\ 0.0332400 & -0.2326400 & -0.9273100 & -0.7212400 \end{bmatrix}
$$

The eigenvalues found in this example are $(-0.360637 \pm 4.806163\,i)$ and $(-0.022822 \pm 1.173937\,i)$, where $i = \sqrt{-1}$. The same results are reported by Khot (1994) but up to 4 decimal points only.

4.5.2 Example 2

This example is a 24×24 real unsymmetric open-loop matrix **A** resulting from the tetrahedral truss shown in Figure 4.4 and solved by Khot (1994). The truss has 12 members and 24 state variables (3 displacements and 3 velocities at nodes 1 to 4). Six actuators and six sensors are collocated along members 7 to 12. A lumped mass of 2 units is attached to nodes 1 to 4. The modulus of elasticity of the members and the unit weight of the structural material are taken as 1.0 and 0.001 in unspecified units respectively. The weighting 24×24 **Q** and 6×6 **R** matrices are chosen as identity matrix. The eigenvalues found in this example are

$(0.0 \pm 12.905109\,i)$, $(0.0 \pm 10.284774\,i)$, $(0.0 \pm 9.250564\,i)$,
$(0.0 \pm 8.539417\,i)$, $(0.0 \pm 4.755259\,i)$, $(0.0 \pm 4.662068\,i)$,
$(0.0 \pm 4.204482\,i)$, $(0.0 \pm 3.398199\,i)$, $(0.0 \pm 2.957414\,i)$,

$(0.0 \pm 2.890711 \, i \,)$, $(0.0 \pm 1.664723 \, i \,)$, $(0.0 \pm 1.342011 \, i \,)$.

The same results are reported by Khot (1994) but up to 4 decimal points only.

4.5.3 Examples 3, 4 and 5

These examples are three randomly generated 100×100, 200×200, and 500×500 real unsymmetric matrices, respectively.

4.5.4 Examples 6

This example is a 1632×1632 real unsymmetric matrix **W** defined by Eq. (4.1) for the 21-story space truss rotated pentagon hollow structure shown in Figures 4.5 and 4.6. The structure has 560 members and 136 nodes. This results in 816 state variables (3 displacements and 3 velocities for each node of the structure). Sixty-five actuators and sixty five sensors are collocated along the horizontal members in the upper level (section II) of the structure. The weighting 816×816 **Q** and 65×65 **R** matrices are chosen as identity matrix.

4.6 PERFORMANCE RESULTS

The vectorization performance of the parallel-vector algorithms in terms of MFLOPS (million floating point operations per second) is shown in Table 4.7. The FLOP (floating point operations) and MFLOPS numbers were found using the Cray system function *hpm* (for hardware performance monitors). The vectorization performance increases significantly with the size of the problem. The MFLOPS for the largest problem resulting from the

21-story structure is a high 204.42. This is significant in the light of highly iterative nature of the complex eigenvalue problem.

Table 4.7: Vectorization performance of the parallel-vector algorithms in terms of MFLOPS.

Example	Matrix size	Number of FLOP	MFLOPS
Example 3	100×100	1×10^7	111.10
Example 4	200×200	9×10^7	130.00
Example 5	500×500	1.4×10^9	150.00
Example 6	1632×1632	5×10^{10}	204.42

Speedups resulting from vectorization on a single processor for examples 3 to 6 are shown in Figure 4.7. This figure shows clearly that the vectorization speedup increases with the size of the problem. Speedups due to parallel processing only (without the use of vectorization) are presented in Figure 4.8. While Cray YMP 8E/8128 has eight processors we had simultaneous access only to seven processors. Finally, Figure 4.9 shows the speedups due to combined parallel processing and vectorization.

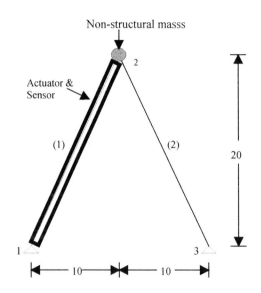

Figure 4.3: Example 1 - Two-bar truss

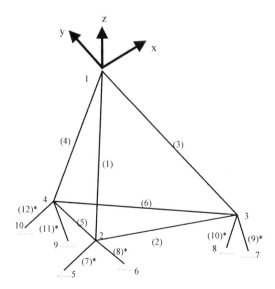

* An actuator exists along the member

**Figure 4.4: Example 2 - Tetrahedral truss (members are
identified in parentheses).**

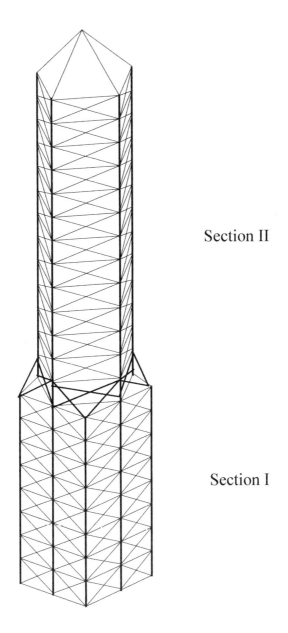

Section II

Section I

Figure 4.5: Example 6 - 21-story space truss structure.

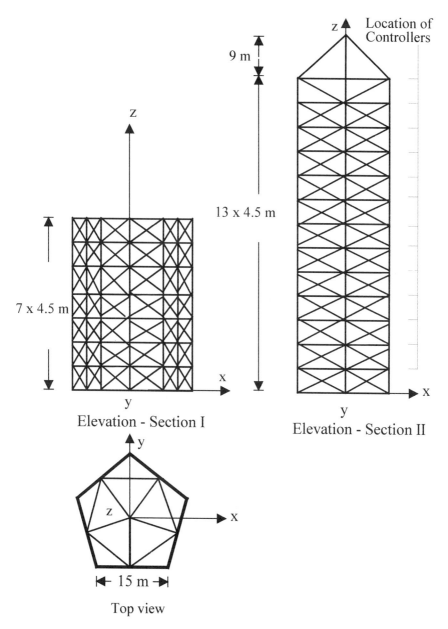

9 m

z

13 x 4.5 m

7 x 4.5 m

z

x

y

Elevation - Section I

Location of
Controllers

z

x

y

Elevation - Section II

y

z

x

|← 15 m →|

Top view

Figure 4.6: Top view and elevations of the structure in Figure 4.5

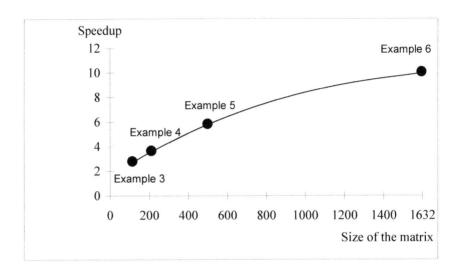

Figure 4.7: Speedup resulting from vectorization on a single processor for examples 3 to 6

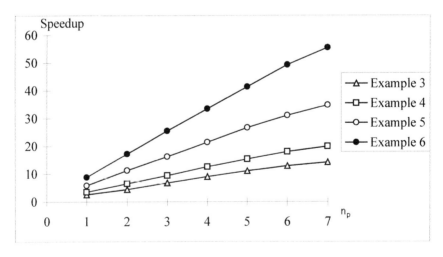

Figure 4.8: Speedups due to parallel processing for examples 3 to 6(without the use of vectorization)

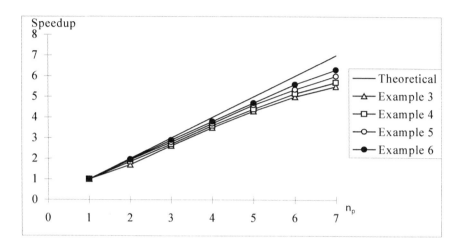

Figure 4.9: Speedups due to combined parallel processing and vectorization for examples 3 to 6

For the largest example, the speedup due to vectorization alone is 9.8, due to parallel processing is 6.3 using seven processors, and due to simultaneous application of vectorization and parallel processing is a very significant 52.4 (using seven processors).

4.7 CONCLUSIONS

Parallel-vector algorithms were presented for the solution of the complex eigenvalue problem encountered in the integrated structural/control optimization problem. The accuracy of the algorithms was verified by comparison with existing solutions in the literature for small problems as well as sequential software packages EISPACK (Garbow et al., 1977) and LINPACK (Stewart et al., 1979). These packages have been developed for use on uniprocessor machines only and thus are limited in handling large complex eigenvalue problems.

The algorithms presented in this book are insensitive to initial values for the eigenvectors. After trying various initial values we found the algorithms converge to the same accurate solutions within 2 (for good initial estimates) to 5 (for poor initial estimates) iterations. The selection of initial values has an effect on the load assigned to each processor and consequently the total execution time. But, it has no effect on the parallel processing speedup because the algorithms are created such that each processor performs the same number of iterations.

The algorithms provide stable results consistently for problems of various size and can find all the eigenvalues of a general real matrix including repeated and closely spaced ones without any numerical difficulty.

The vectorization and parallel processing efficiencies of the parallel-vector algorithms were investigated by application to large problems. It was shown that they are particularly efficient for large problems.

CHAPTER 5

PARALLEL ALGORITHMS FOR SOLUTION OF THE RICCATI EQUATION

5.1 INTRODUCTION

One of the most frequently used nonlinear equations in active control of structures is the Riccati equation (Eq. 3.15). The Riccati equation is a class of matrix quadratic algebraic equations arising in the study of controlled continuous or discrete dynamic systems. The Riccati equation also arrises in many other applications in optimal control such as the control of turboprop engines, boilers, nuclear reactors, and aircrafts (Morse and Wonham, 1971, and Friedland, 1986) and in filtering and prediction of noise in applications such as flow rate variations in chemical reactors, wind gust, and updrafts acting on aircrafts and rockets (Johnson, 1971, Rhodes, 1971, and Friedland, 1986).

The solution of the Riccati equation is the most time-consuming part of any optimal control problem. It requires an inordinate amount of processing time when applied to large problems. Thus, development of parallel-vector algorithms for the solution of the Riccati equation, using the vector and multi-processing capabilities of high performance computers (Adeli, 1992a&b) provides an opportunity to solve large problems never solved before.

In the following section, various approaches for the solution of the Riccati equation are first reviewed in terms of their suitability for parallel processing and stability for solving large systems. Then, the approach used in this book is presented to be followed by presentation of parallel-vector algorithms. Finally, several examples are presented and performance results are discussed.

5.2 APPROACHES FOR THE SOLUTION OF THE RICCATI EQUATION

A number of different methods have been proposed to solve the algebraic Riccati equation. One of the early approaches is the eigenvector method (Potter, 1966, Arnold and Laub, 1984, and Meirovitch, 1990). In this method, first a $4N \times 4N$ Hamiltonian matrix, $\mathbf{W}$, is formed[1]

$$\mathbf{W} = \begin{bmatrix} \mathbf{A}^T & \mathbf{Q} \\ \mathbf{BR}^{-1}\mathbf{B}^T & -\mathbf{A} \end{bmatrix} \qquad (repeated) \qquad (4.1)$$

Then, the complex eigenvalue problem of matrix $\mathbf{W}$ is solved and its eigenvectors are used to calculate the Riccati matrix, $\mathbf{P}$. The eigenvector approach is known to yield accurate results (Potter, 1966, and Meirovitch, 1990). But, it requires an inordinate amount of processing time, mostly for the solution of the complex eigenvalue problem of matrix $\mathbf{W}$.

A variant of the eigenvector method is the Schur method (Laub, 1979, Paige and Loan, 1981). In this method, a $4N \times 4N$ Hamiltonian matrix, $\overline{\mathbf{W}}$, is formed ($\overline{\mathbf{W}}$ has conjugate complex eigenvalues)

---

[1]   Matrix $\mathbf{W}$ is Hamiltonian if $-\mathbf{W} = \mathbf{J}^{-1}\mathbf{W}^T\mathbf{J}$, where

$$\mathbf{J} = \begin{bmatrix} 0 & \mathbf{I}_{2N} \\ -\mathbf{I}_{2N} & 0 \end{bmatrix}$$ and $\mathbf{I}_{2N}$ is an identity matrix of order $2N$.

$$\overline{\mathbf{W}} = \begin{bmatrix} \mathbf{A} & -\mathbf{BR}^{-1}\mathbf{B}^T \\ -\mathbf{Q} & -\mathbf{A}^T \end{bmatrix} \tag{5.1}$$

Then, in each iteration s, a $4N \times 4N$ matrix, $\overline{\mathbf{W}}^{(s)}$, (a superscript in parentheses refers to the iteration number) is obtained from the following orthogonal transformation:

$$\overline{\mathbf{W}}^{(s)} = \mathbf{U}^{(s)^T} \overline{\mathbf{W}} \mathbf{U}^{(s)} \tag{5.2}$$

where $\mathbf{U}^{(s)}$ is a $4N \times 4N$ transformation matrix (Laub 1979). Then, in the next iteration, $s+1$, a new matrix $\overline{\mathbf{W}}^{(s+1)}$ is obtained as follows:

$$\overline{\mathbf{W}}^{(s+1)} = \mathbf{U}^{(s+1)^T} \overline{\mathbf{W}}^{(s)} \mathbf{U}^{(s+1)} \tag{5.3}$$

This iterative process is continued until an upper triangular matrix, $\overline{\mathbf{W}}^{(s+t)}$, is obtained:

$$\overline{\mathbf{W}}^{(s+t)} = \begin{bmatrix} \overline{\mathbf{W}}_{11}^{(s+t)} & \overline{\mathbf{W}}_{12}^{(s+t)} \\ 0 & \overline{\mathbf{W}}_{22}^{(s+t)} \end{bmatrix} \tag{5.4}$$

where $(s+t)$ is the final number of iterations. The real parts of the $2N$ complex eigenvalues of the real matrix $\overline{\mathbf{W}}_{11}^{(s+t)}$ are negative and the real parts of the $2N$ complex eigenvalues of the real matrix $\overline{\mathbf{W}}_{22}^{(s+t)}$ are positive. Then, a $4N \times 4N$ matrix is found as follows:

$$\mathbf{U} = \begin{bmatrix} \mathbf{U}_{11} & \mathbf{U}_{12} \\ \mathbf{U}_{21} & \mathbf{U}_{22} \end{bmatrix} = \mathbf{U}^{(s+t)}\mathbf{U}^{(s+t-1)}\mathbf{U}^{(s+t-2)}\ldots\mathbf{U}^{(2)}\mathbf{U}^{(1)} \quad (5.5)$$

in which $\mathbf{U}_{ij}$'s are $2N \times 2N$ real submatrices and $\mathbf{U}_{11}$ is invertible. The Riccati matrix is finally obtained from the following equation (Laub 1979):

$$\mathbf{P} = \mathbf{U}_{21}\mathbf{U}_{11}^{-1} \qquad (5.6)$$

The Schur method is known to be numerically stable and reliable for systems with matrices of size up to 100 (Laub, 1979). This method, however, is not amenable for effective parallel processing because of the interdependence of the transformations in each iteration.

Another approach is the matrix sign function method (Gardiner and Laub, 1986, Byers, 1987). If matrix $\mathbf{W}$ can be transformed into the Jordan canonical form[2]

$$\left(\overline{\mathbf{D}} + \overline{\overline{\mathbf{D}}}\right) = \overline{\mathbf{U}}^{-1}\mathbf{W}\overline{\mathbf{U}} \qquad (5.7)$$

where $\overline{\mathbf{U}}$ is a $4N \times 4N$ transformation matrix, $\overline{\mathbf{D}}$ is a $4N \times 4N$ diagonal matrix, and $\overline{\overline{\mathbf{D}}}$ is a $4N \times 4N$ matrix that commutes with matrix $\overline{\mathbf{D}}$ ($\overline{\mathbf{D}}\,\overline{\overline{\mathbf{D}}} = \overline{\overline{\mathbf{D}}}\,\overline{\mathbf{D}}$). The $4N \times 4N$ matrix $\mathrm{SIGN}(\mathbf{W})$ is then defined by (Byers, 1987)

$$\mathrm{SIGN}(\mathbf{W}) = \overline{\mathbf{U}}\mathbf{S}\overline{\mathbf{U}}^{-1} \qquad (5.8)$$

---

[2]   In this form, the matrix consists of submatrices on the diagonal where each submatrix is an upper triangular matrix.

in which $\mathbf{S}$ is a $4N \times 4N$ diagonal matrix whose diagonal elements S_{ii} are given by

$$S_{ii} = \begin{cases} 1 & \text{if } \operatorname{Re}(\overline{d}_{ii}) > 0 \\ -1 & \text{if } \operatorname{Re}(\overline{d}_{ii}) < 0 \end{cases} \tag{5.9}$$

where $\operatorname{Re}(d_{ii})$ denotes the real part of the complex number $\overline{d}_{ii}$, the diagonal term of $\overline{\mathbf{D}}$. If $\mathbf{W}$ has any pure imaginary eigenvalue (with no real part), $\operatorname{SIGN}(\mathbf{W})$ is not defined and the method fails. For the Hamiltonian matrix $\mathbf{W}$ in Eq. (5.1), the matrix $\operatorname{SIGN}(\mathbf{W})$ is given by (Byers, 1987)

$$\operatorname{SIGN}(\mathbf{W}) = \begin{bmatrix} \mathbf{P} & -\mathbf{I}_{2N} \\ \mathbf{I}_{2N} & \mathbf{0}_{2N} \end{bmatrix} \begin{bmatrix} \mathbf{I}_{2N} \\ \mathbf{0}_{2N} \end{bmatrix} \begin{bmatrix} \mathbf{Z} \\ -\mathbf{I}_{2N} \end{bmatrix} \begin{bmatrix} \mathbf{P} & -\mathbf{I}_{2N} \\ \mathbf{I}_{2N} & \mathbf{0}_{2N} \end{bmatrix}^{-1} \tag{5.10}$$

where $\mathbf{0}_{2N}$ is a $2N \times 2N$ zero matrix and $\mathbf{Z}$ is a $4N \times 4N$ matrix to be defined later. In this method, in the first iteration $s = 1$, a $4N \times 4N$ matrix $\hat{\mathbf{W}}^{(s)}$, is set equal to $\mathbf{W}$ given by Eq. (5.1) and the Riccati matrix, $\mathbf{P}$, is obtained from the solution of the equation

$$\mathbf{FP} = -\mathbf{E} \tag{5.11}$$

where $\mathbf{F}$ is a $4N \times 2N$ matrix representing the first $2N$ columns of the matrix $\hat{\mathbf{W}}^{(s)} - \mathbf{I}_{4N}$, $\mathbf{E}$ is a $4N \times 2N$ matrix representing the last $2N$ columns of the matrix $\hat{\mathbf{W}}^{(s)} - \mathbf{I}_{4N}$, and $\mathbf{I}_{4N}$ is a $4N \times 4N$ identity matrix. Next, in each iteration, $s+1$, the $4N \times 4N$ matrices $\mathbf{Z}^{(s+1)}$ and $\hat{\mathbf{W}}^{(s+1)}$ are obtained as follows:

$$\mathbf{Z}^{(s+1)} = \hat{\mathbf{W}}^{(s)} \left| \det\left[\hat{\mathbf{W}}^{(s)} \right] \right|^{\frac{-1}{2N}} \tag{5.12}$$

$$\hat{\mathbf{W}}^{(s+1)} = \mathbf{Z}^{(s+1)} - \left[\frac{\mathbf{Z}^{(s+1)} - \left[\mathbf{JZ}^{(s+1)} \right]^{-1} \mathbf{J}}{2} \right] \tag{5.13}$$

A new Riccati matrix $\hat{\mathbf{P}}$ is then obtained from the solution of Eq. (5.11) using the updated $\hat{\mathbf{W}}^{(s+1)}$. The iterations continue until $\hat{\mathbf{P}}$ approaches $\mathbf{P}$. Although this method seems to be suitable for parallel computations, it can be numerically unstable (Gardiner, 1997) and convergence cannot be guatanteed for large problems. Gardiner (1997) attempted to improve the stability of the method using a combination of iterative refinement, scaling, and shifting techniques and presented results for matrices of order 100. We tried the stabilized matrix sign function algorithm for the solution of the different examples presented in this chapter. The algorithm produced stable results for two small examples reported in the literature but failed to converge for Examples 1 to 3 to be presented in this chapter subsequently.

A fourth approach is the Newton's iterative method (Kleinman, 1968, Hewer, 1971, and Sandell, 1974). In this method, the Riccati matrix $\mathbf{P}$ is initially selected as identity matrix. Then, in each iteration s, a sequence of matrices is computed as follows:

$$\hat{\mathbf{F}}^{(s)} = \left[\mathbf{R} + \mathbf{B}^T \mathbf{P}^{(s-1)} \mathbf{B} \right]^{-1} \mathbf{B}^T \mathbf{P}^{(s-1)} \mathbf{A} \tag{5.14}$$

$$\hat{\mathbf{E}}^{(s)} = \mathbf{A} - \mathbf{B}\hat{\mathbf{F}}^{(s)} \tag{5.15}$$

$$\mathbf{P}^{(s)} = \hat{\mathbf{E}}^{(s)^T} \mathbf{P}^{(s)} \hat{\mathbf{E}}^{(s)} + \hat{\mathbf{F}}^{(s)^T} \mathbf{R}\hat{\mathbf{F}}^{(s)} + \mathbf{Q} \qquad (5.16)$$

This iterative process is continued until convergence is achieved. This method is implemented in FORTRAN in the ORACLE software package for solution of linear quadratic-Guasian control problems (Armstrong, 1980) on uniprocessor machines. The routine can be applied only to small matrices of size less than 100 (Armstrong, 1978). Computational difficulties are encountered in the calculation of the inverse of the matrix $(\mathbf{R}+\mathbf{B}^T \mathbf{PB})$ for large problems (order of several hundreds).

5.3 THE METHOD USED IN THIS WORK

Out of the four methods described in the previous section, we found the eigenvector approach to be the most accurate and stable method especially for large problems. The method also lends itself to effective parallel processing if efficient parallel algorithms can be developed for the solution of the complex eigenvalue problem of a general real unsymmetric matrix. In Chapter 4, we presented efficient parallel-vector algorithms for the solution of the complex eigenvalue problem. In this chapter, we present robust parallel algorithms for the solution of the Riccati equation.

We start with forming the Hamiltonian matrix $\mathbf{W}$, Eq. (5.1). Then, the eigenvalue problem of matrix $\mathbf{W}$ is solved. The matrix $\mathbf{W}$ has 2N complex eigenvalues with positive real parts and 2N complex eigenvalues with negative real parts. The eigenvectors associated with the complex eigenvalues that have positive real parts are used to find the solution to the Riccati equation. The Riccati matrix $\mathbf{P}$ is obtained from the solution of the 2N systems of 2N linear equations (Meirovitch, 1990)

$$\mathbf{P} = \hat{\mathbf{E}}\hat{\mathbf{F}}^{-1} \tag{5.17}$$

where the $2N \times 2N$ matrices, $\hat{\mathbf{E}}$ and $\hat{\mathbf{F}}$, are the upper and lower halves of the eigenvectors associated with the complex eigenvalues with positive real parts. The approach chosen for the solution of Eq. (5.17) depends on whether $\mathbf{P}$ is positive definite, as explained in the next section.

5.4 PARALLEL-VECTOR ALGORITHMS

The parallel-vector algorithms for the solution of the Riccati equation are presented in Figures 5.1a to d and Table 5.1.

A major portion of the computation time is consumed by the solution of the complex eigenvalues of the Hamiltonian matrix $\mathbf{W}$, Eq. (5.1). In Chapter 5, we presented efficient parallel-vector algorithms for the solution of the complex eigenvalue problem by minimizing the processing time consumed by bottleneck operations that degrade the parallel processing performance such as inner product operations and the solution of systems of linear equations frequently encountered in the process.

Another time consuming part in the solution of the Riccati equation is the solution of the $2N$ systems of $2N$ linear equations, Eq. (5.17), to compute the Riccati matrix $\mathbf{P}$. The parallel processing approach used for the solution of the Riccati matrix $\mathbf{P}$ depends on whether this matrix is positive definite or not. When $\mathbf{P}$ is not positive definite, the $2N$ systems of $2N$ linear equations are distributed among n_p processors as evenly as possible using macrotasking. Each system of $2N$ linear equations is then solved on a separate processor using vector operations (Figures 5.1a to c).

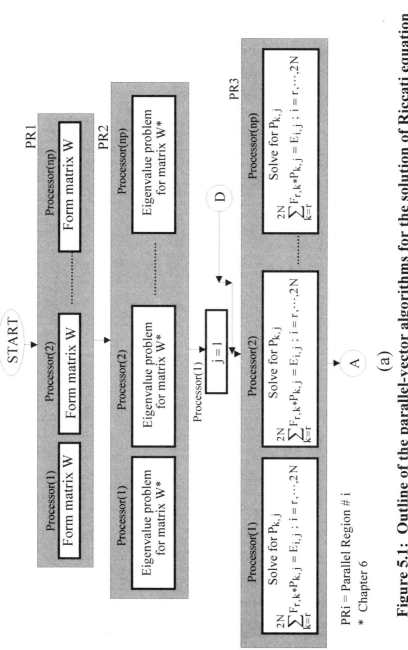

Figure 5.1: Outline of the parallel-vector algorithms for the solution of Riccati equation

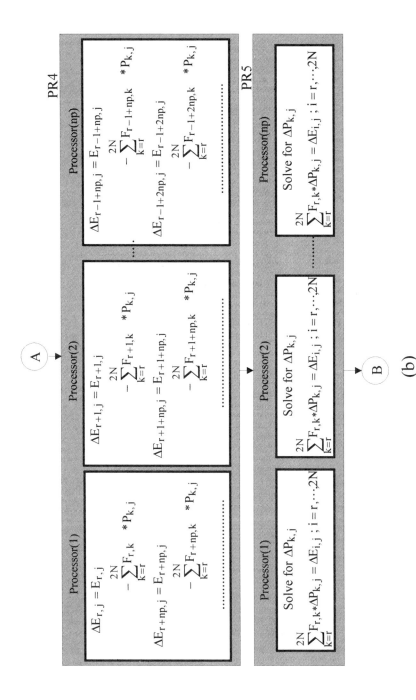

Figure 5.1 - continued

(b)

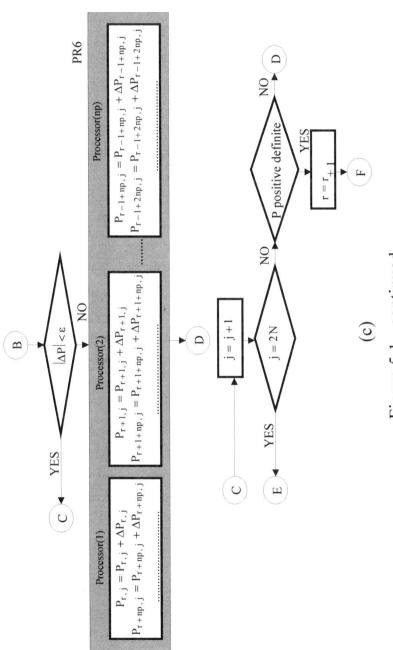

Figure 5.1 - continued

(c)

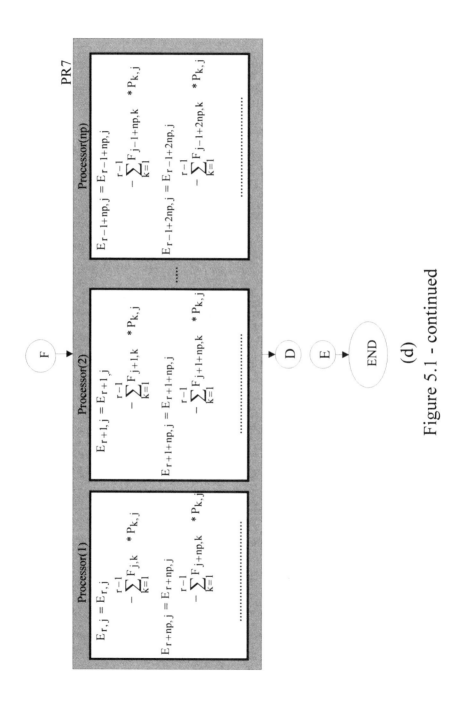

Figure 5.1 - continued

(d)

Table 5.1   Parallel-vector algorithms for the solution of the Riccati equation

Input: N, N_{f_o}, **A, B, Q, R,** n_p, ε, $\lambda_1 \ldots \lambda_{4N}$, $\mathbf{e}_1 \ldots \mathbf{e}_{4N}$

$\mathbf{D} = \mathbf{BR}^{-1}\mathbf{B}^T$ (**Microtasking and Vectorization**)

FOR $p = 1$ **UNTIL** n_p **DO**   (**Microtasking**)

$\quad i = p$

(a)   **FOR** $j = 1$ **UNTIL** $2N$ **DO**   (**Vectorization**)

$\quad\quad \mathbf{W}_{i,j} = \mathbf{A}_{j,i}$

$\quad\quad \mathbf{W}_{i,j+2N} = \mathbf{Q}_{i,j}$

$\quad\quad \mathbf{W}_{i+2N,j} = \mathbf{D}_{i,j}$

$\quad\quad \mathbf{W}_{i+2N,j+2N} = -\mathbf{A}_{i,j}$

$\quad$ **NEXT** j

$\quad i = i + n_p$

$\quad$ **IF** $i \leq 2N$ **THEN** Go to (a)

NEXT p

Solve the eigenvalue problem of matrix **W** *using the parallel-vector algorithms presented in Tables 4.1 to 4.3.*

FOR $p = 1$ **UNTIL** n_p **DO**   (**Microtasking**)

$\quad i = p$

(b)   **IF** Re $(\lambda_i > 0)$ **THEN**

$\quad\quad$ **FOR** $j = 1$ **UNTIL** $2N$ **DO**   (**Vectorization**)

$\quad\quad\quad \mathbf{E}_{i,j} = \mathbf{e}_{i_j}$

$\quad\quad\quad \mathbf{F}_{i,j} = \mathbf{e}_{i_{j+2N}}$

$\quad\quad$ **NEXT** j

$\quad i = i + n_p$

Table 5.1 – continued

IF $i \le 2N$ **THEN** Go to (b)
NEXT p
$j = 1, r = 1$

(c)
$$\begin{bmatrix} \mathbf{P}_{r,j} \\ \cdots \\ \mathbf{P}_{2N,j} \end{bmatrix} = \begin{bmatrix} \mathbf{F}_{r,r} & \cdots & \mathbf{F}_{r,2N} \\ \cdots & \cdots & \cdots \\ \mathbf{F}_{2N,r} & \cdots & \mathbf{F}_{2N,2N} \end{bmatrix}^{-1} \begin{bmatrix} \mathbf{E}_{r,j} \\ \cdots \\ \mathbf{E}_{2N,j} \end{bmatrix}$$

(Microtasking and Vectorization)
(d)   **FOR** $p = 1$ **UNTIL** n_p **DO**   **(Microtasking)**
       $i = p\text{-}1$
(e)   **FOR** $k = j$ **UNTIL** $2N$ **DO**   **(Vectorization)**
         $\Delta \mathbf{E}_{r+i,j} = \mathbf{E}_{r+i,j} - \mathbf{F}_{r+i,k} \ast \mathbf{P}_{k,j}$
       NEXT k
       $i = i + n_p$
       IF $i \le 2N$ **THEN** Go to (e)
       NEXT p

$$\begin{bmatrix} \Delta\mathbf{P}_{r,j} \\ \cdots \\ \Delta\mathbf{P}_{2N,j} \end{bmatrix} = \begin{bmatrix} \mathbf{F}_{r,r} & \cdots & \mathbf{F}_{r,2N} \\ \cdots & \cdots & \cdots \\ \mathbf{F}_{2N,r} & \cdots & \mathbf{F}_{2N,2N} \end{bmatrix}^{-1} \begin{bmatrix} \Delta\mathbf{E}_{r,j} \\ \cdots \\ \Delta\mathbf{E}_{2N,j} \end{bmatrix}$$

(Microtasking and Vectorization)
IF $|\Delta\mathbf{P}| < \varepsilon$ **THEN**
       $j = j + 1$
       IF $j = 2N$ **THEN** Go to (i)

Table 5.1 – continued

ELSE

    IF **P** is positive definite **THEN**

        $r = r + 1$

        Go to (g)

    ELSE Go to (c)

ELSE

    Continue

FOR $p = 1$ **UNTIL** n_p **DO**   (**Microtasking**)

    $i = p - 1 + r$

(f) $\mathbf{P}_{i,j} = \mathbf{P}_{i,j} + \Delta\mathbf{P}_{i,j}$ (**Vectorization**)

    $i = i + n_p$

    IF $i \leq 2N$ **THEN** Go to (f)

NEXT p

Go to (d)

(g)   **FOR** $p = 1$ **UNTIL** n_p **DO**

      $i = p - 1$

(h)   **FOR** $k = 1$ **UNTIL** r-1 **DO**   (**Vectorization**)

          $\mathbf{E}_{r+i,j} = \mathbf{E}_{r+i,j} - \mathbf{F}_{j+i,k} \quad *\,\mathbf{P}_{k,j}$

      NEXT k

      $i = i + n_p$

      IF $i \leq 2N$ **THEN** Go to (h)

   NEXT p

   Go to (c)

(i)   **STOP**

The total number of unknowns in the Riccati matrix $\mathbf{P}$ in this case is $4N^2$. When $\mathbf{P}$ is positive definite (which is the case in structural control applications) the symmetry of this matrix is considered and the total number of unknowns in the Riccati matrix is reduced to $2N^2 + N$. In this case, we first solve a system of $2N$ linear equations concurrently using microtasking as well as vectorization. In the next step, we solve a system of $2N$ - 1 linear equations concurrently. This process is continued until we find all the unknowns, where in each step the number of linear equations solved is one fewer than the number of equations solved in the previous step.

The parallel-vector algorithms outlined in Figures 5.1a to d consist of 7 parallel regions to be described shortly. Between these parallel regions there exists some operations that are performed by one processor, called master processor and identified as processor (1) in Figure 5.1. These operations which are kept at a minimum include the summation in the inner product operation, convergence check, and other checks. A short description of each parallel region follows:

PR1: formation of matrix $\mathbf{W}$ (Figure 5.1a).

PR2: solution of the eigenvalue problem for matrix $\mathbf{W}$ (Figure 5.1a).

PR3: computation of the jth column of the Riccati matrix $\mathbf{P}$ (Figure 5.1a).

PR4: computation of the error in the solution of the jth column of matrix $\mathbf{P}$ (Figure 5.1b).

PR5: computation of the increment for the jth column of improved matrix $\mathbf{P}$ (Figure 5.1b).

PR6: computation of the jth column of improved matrix $\mathbf{P}$ (Figure 5.1c).

PR7: computation of the system of $2N$ - j linear equations (Figure 5.1d).

5.5 EXAMPLES

In order to verify the accuracy of the parallel-vector algorithms presented in this chapter, we first used two small examples reported in the recent literature. The first example is a two-bar truss solved by Khot (1994) with one actuator and one sensor (Figure 4.3). For this example, the Hamiltonian matrix **W** is 8×8 and the Riccati matrix **P** is 4×4. The second example is a tetrahedral truss with 12 members and 24 state variables solved by Khot (1994) (Figure 4.4) who reported results up to four decimal points. We obtained the same results for both examples.

Next, three large examples were created with the objective of demonstrating the robustness and efficiency of the parallel-vector algorithms presented in this chapter. Efficiency is presented in terms of MFLOPS and the speedup.

5.5.1 Example 1

The Riccati equation for the two-span continuous truss bridge structure shown in Figure 5.2 is solved. The structure has 388 members and 100 nodes. This results in 564 state variables (3 displacements and 3 velocities for each node of the structure). Fifty six actuators and fifty six sensors are collocated along the members in the vertical planes in the middle half of each span of the structure. The weighting 564×564 **Q** and 56×56 **R** matrices are chosen as identity matrix. The complex eigenvalue problem of the unsymmetric 1128×1128 Hamiltonian matrix **W** is solved and the 564×564 Riccati matrix **P** is obtained. The matrix sign function algorithm (Gardiner, 1997) failed to produce stable results for this example.

5.5.2 Example 2

The Riccati equation for the 21-story space truss pentagon hollow structure shown in Figure 5.3 is solved. The structure has 205 members and 101 nodes. This results in 582 state variables (3 displacements and 3 velocities for each node of the structure). Sixty actuators and sixty sensors are collocated along the horizontal members in floors 10 to 20 and in the five members of the 21st floor (Figure 5.3). The weighting 582×582 **Q** and 60×60 **R** matrices are chosen as identity matrix. The complex eigenvalue problem of the 1164×1164 Hamiltonian matrix **W** is solved and the 582×582 Riccati matrix **P** is obtained. The matrix sign function algorithm (Gardiner, 1997) failed to produce stable results for this example.

5.5.3 Example 3

The Riccati equation for the 12-story moment-resisting space frame structure shown in Figure 5.4 is solved. The rotated-square structure has 152 members and 77 nodes. This results in 816 state variables (3 displacements, 3 rotations, 3 linear velocities, and 3 angular velocities for each node of the structure). Thirty two actuators and thirty two sensors are collocated along the horizontal members in the upper eight stories (section II) of the structure. The weighting 816×816 **Q** and 28×28 **R** matrices are chosen as identity matrix. The complex eigenvalue problem of the 1632×1632 Hamiltonian matrix **W** is solved and the 816×816 Riccati matrix **P** is obtained. The matrix sign function algorithm (Gardiner, 1997)

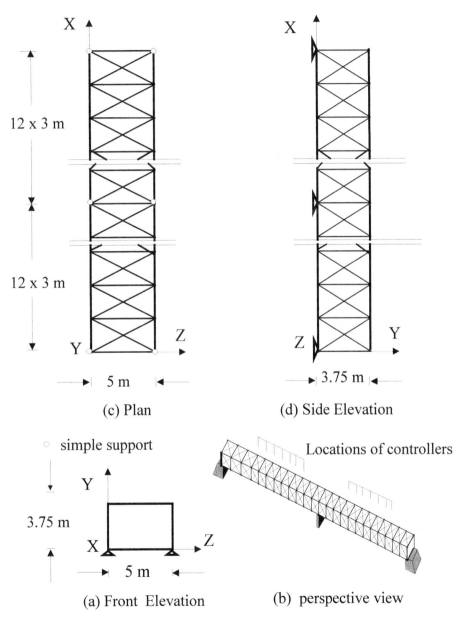

(c) Plan

(d) Side Elevation

(a) Front Elevation

(b) perspective view

Figure 5.2: Example 1 - Two-span space truss structure

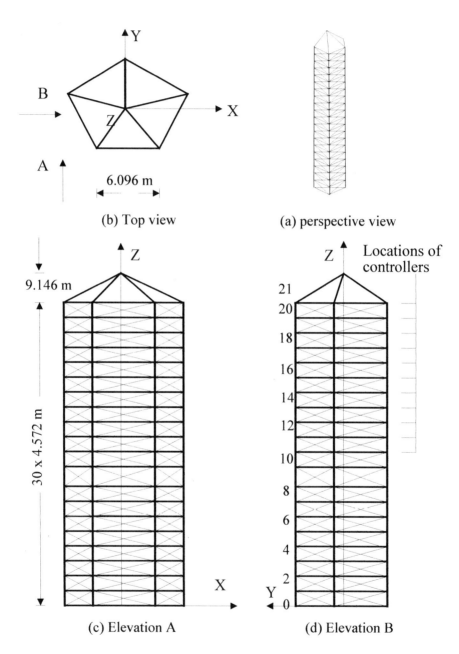

Figure 5.3: Example 2 - 21-story space truss structure

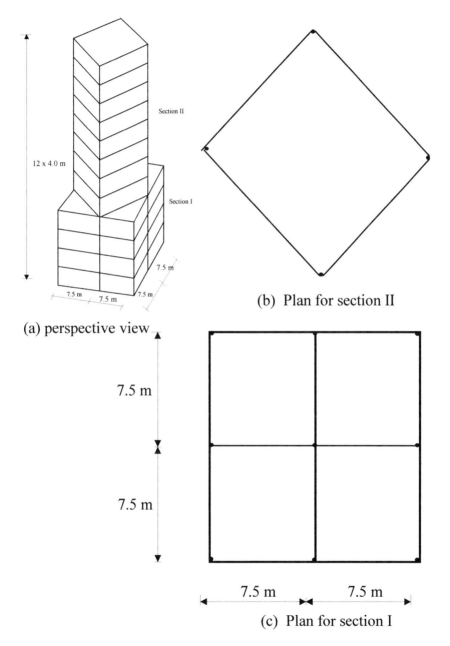

(a) perspective view

(b) Plan for section II

(c) Plan for section I

Figure 5.4: Example 3 - 12-story moment-resisting space frame structure

failed to produce stable results for this example.

5.6 PERFORMANCE RESULTS

The vectorization performance of the parallel-vector algorithms in terms of MFLOPS is shown in Table 5.2. The vectorization performance increases with the size of the problem. The MFLOPS for the largest problem resulting from the 12-story space frame structure is a high 206.0.

Speedups resulting from vectorization on a single processor for examples 1 to 3 are shown in Figure 5.5. This figure shows clearly the vectorization speedup increases with the size of the problem. Speedups due to parallel processing alone (without the use of vectorization) are presented in Figure 5.6. Finally, Figure 5.7 shows the speedups due to combined parallel processing and vectorization.

Table 5.2: Performance results

Example	Size of matrix **P**	MFLOPS	Total CPU time using vectorization and parallel processing with 7 processors (Sec.)
Example 1	564×564	178.6	266.32
Example 2	606×606	182.7	405.57
Example 3	816×816	206.0	1080.8

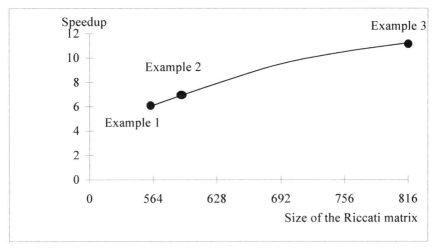

**Figure 5.5:   Speedup resulting from vectorization on a single
            processor for examples 1 to 3**

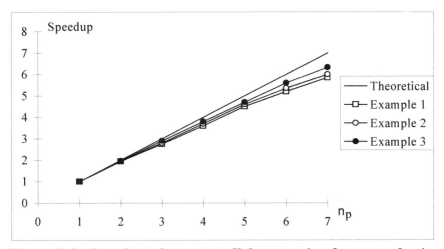

**Figure 5.6:   Speedups due to parallel processing for examples 1
            to 3 (without the use of vectorization)**

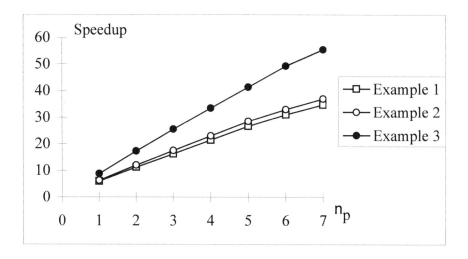

**Figure 5.7: Speedups due to simultaneous parallel processing
and vectorization for examples 1 to 3**

For the largest example (example 3), the speedup due to vectorization alone is 11.2, due to parallel processing using seven processors is 6.33, and due to simultaneous application of vectorization and parallel processing is a very significant 54.4 (using seven processors). The total CPU time using 7 processors with vectorization for the largest example (Example 3) is 1080.8 seconds.

5.7 CONCLUSIONS

Efficient and robust parallel-vector algorithms were presented for the solution of the Riccati equation encountered in optimal control problems. The accuracy of the algorithms was verified by comparison with solutions reported by other researchers in the literature for small problems. Further, the algorithms were applied to three large examples created in this work.

The algorithms provide stable results consistently for problems of various size while other algorithms show numerical instability for large problems. The vectorization and parallel processing efficiencies of the parallel-vector algorithms were investigated by application to large problems. It is shown that the efficiency increases with the size of the problem.

CHAPTER 6

SMART BRIDGE STRUCTURES

6.1 INTRODUCTION

In this chapter, we present a computational model for active control of large structures subjected to dynamic loadings such as impact, wind or earthquake loadings. We use actuators collocated at member ends and distributed throughout the structure. The governing differential equations of the open loop and closed loop systems are formulated and a recursive approach to compute the response of the structure is presented. A robust parallel-vector algorithm is presented for the recursive solution of the response of the open loop and closed loop systems. For large structures, the major bottleneck in this problem is the solution of the complex eigenvalue problem encountered in the solution of the resulting Riccati equation as well as the solution of both open loop and closed loop systems of equations. The methods reported in the literature yield satisfactory results for small problems but often become unstable for large problems (Gardiner, 1997). In Chapter 4, we presented robust parallel algorithms for solution of the complex eigenvalue problem of an unsymmetric matrix and applied them successfully to large matrices.

The computational model is applied to bridge structures in this chapter and to highrise building structures in the following chapter. We investigate three different schemes for placement of controllers in bridge structures through application to active control of three large bridge structures.

6.2 RESPONSE OF CONTROLLED STRUCTURES

As described in Chapter 3, the discretized differential equation governing the uncontrolled motion of a structure, the open loop system of equations, is given by (see Chapter 3 for definition of terms)

$$\dot{\mathbf{X}} = \mathbf{A}\mathbf{X} + \mathbf{B}_\circ \mathbf{f}_\circ \qquad\qquad (\textit{repeated}) \qquad (3.16)$$

The equation governing the controlled motion of a structure, the closed-loop system of equations, is given by

$$\dot{\mathbf{X}} = \overline{\mathbf{A}}\mathbf{X} + \mathbf{B}_\circ \mathbf{f}_\circ \qquad\qquad (\textit{repeated}) \qquad (3.19)$$

where $\overline{\mathbf{A}}$ is the $2N \times 2N$ unsymmetric closed-loop matrix given by

$$\overline{\mathbf{A}} = \mathbf{A} - \mathbf{B}\mathbf{G}, \qquad\qquad (\textit{repeated}) \qquad (3.18)$$

in which $\mathbf{G}$ is the optimum gain matrix obtained by minimizing the control performance index, J (Eq. 3.12), and is given by

$$\mathbf{G} = \mathbf{R}^{-1}\mathbf{B}^T\mathbf{P}, \qquad\qquad (\textit{repeated}) \qquad (3.14)$$

and $\mathbf{P}$ is a $2N \times 2N$ positive definite matrix called the Riccati matrix obtained from the solution of the following Riccati equation:

$$\mathbf{Q} + \mathbf{P}\mathbf{A} + \mathbf{A}^T\mathbf{P} - \mathbf{P}\mathbf{B}\mathbf{R}^{-1}\mathbf{B}^T\mathbf{P} = 0 \qquad (\textit{repeated}) \qquad (3.15)$$

In this equation, $\mathbf{Q}$ is the $2N \times 2N$ state weighting matrix and $\mathbf{R}$ is the $2N \times 2N$ control weighting matrix. The weighting matrix $\mathbf{Q}$ specifies the relative importance of the various components of the state vector $\mathbf{X}$. For example, if X_1 is of concern while X_2 is not, we may choose a state weighting matrix

$$\mathbf{Q} = \begin{bmatrix} 1 & 0 \\ 0 & 0 \end{bmatrix} \tag{6.1}$$

If both X_1 and X_2 are of equal concern, we may choose a state weighting matrix

$$\mathbf{Q} = \begin{bmatrix} 1 & 0 \\ 0 & 1 \end{bmatrix} \tag{6.2}$$

In this work, we choose the state weighting matrix $\mathbf{Q}$ as $q\mathbf{I}$ where $\mathbf{I}$ is a $2N \times 2N$ identity matrix and q is a weight factor. Similarly, the control weighting matrix $\mathbf{R}$ is chosen as $r\mathbf{I}$, where r is a weight factor. The effects of the weight factor r on the controlled response of the structure are investigated.

The solution for the state variable vector, $\mathbf{X}$, in the closed-loop system, Eq. (3.19), can be expressed as (Casti, 1987)

$$\mathbf{X}^{(k+1)} = \Omega \mathbf{X}^{(k)} + \Gamma \mathbf{f}_\circ{}^{(k)} \qquad k = 1,2,\dots \tag{6.3}$$

where the $2N \times 2N$ matrix Ω is given by

$$\Gamma = \exp\left(\delta T \overline{\mathbf{A}}\right) = \mathbf{e} \, \Theta \, \mathbf{e}'^{T} \tag{6.4}$$

the $2N \times N_{f_\circ}$ matrix Γ is given by

$$\Gamma = \left(\int_0^{\delta T} \exp\left(\delta T \overline{\mathbf{A}}\right) dt \right) \mathbf{B}_\circ = \mathbf{e} \, \Theta \, \mathbf{e}'^{T} \mathbf{B}_\circ \tag{6.5}$$

the $2N \times 2N$ matrices Θ and Δ are given by

$$\Theta = \exp(\delta T \Lambda) \tag{6.6}$$

and

$$\Delta = \int_0^{\delta T} \exp(\delta T \Lambda) \, dt \tag{6.7}$$

In Eq. (6.7) Λ is a diagonal matrix whose nonzero diagonal elements are the eigenvalues of matrix $\overline{\mathbf{A}}$, δT is the time increment, $\mathbf{e}$ is the matrix of right eigenvectors obtained from the solution of $\overline{\mathbf{A}}\mathbf{e} = \Lambda \mathbf{e}$, and $\mathbf{e}'$ is the matrix of left eigenvectors obtained from the solution of $\mathbf{e}'\overline{\mathbf{A}} = \Lambda \mathbf{e}'$.

Matrix Θ can be expressed as

$$\Theta_n(\delta T) = \mathbf{I} + \delta T \Lambda + \frac{\delta T^2}{2!} \Lambda^2 + \cdots + \frac{\delta T^n}{n!} \Lambda^n \tag{6.8}$$

where n is the number of terms used to compute the series expansion (in addition to the identity matrix term). The expression in Eq. (6.8) requires $n(n-1)/2$ diagonal matrix multiplications (Meirovitch, 1985). In order to reduce the number of matrix multiplications, in this work instead of Eq. (6.8) we use a recursive equation to compute Θ which requires only $(n-1)$ matrix multiplications:

$$\Theta_n(\delta T) = \mathbf{I} + \delta T \Lambda \, \zeta_{n-1} \tag{6.9}$$

where ζ_i is a $2N \times 2N$ matrix defined by

$$\zeta_1 = \mathbf{I} + \frac{\delta T \Lambda}{n} \tag{6.10}$$

and

$$\zeta_i = \mathbf{I} + \frac{\delta T \Lambda}{n+1-i} \; \zeta_{i-1}, \qquad i = 2, 3, ..., n. \tag{6.11}$$

The series expansion of Eq. (6.11) will yield Eq. (6.8). Thus, in order to obtain the response of the closed loop structure from Eq. (6.3), we first need to find the complex right and left eigenvectors of the unsymmetric matrix $\overline{\mathbf{A}}$ which is covered in Chapter 4. Then, the response is obtained through the application of the recursive Eqs. (6.9) to (6.11).

The solution for the state variable vector, $\mathbf{X}$, in the open-loop system, Eq. (3.16), is obtained similarly by replacing matrix $\overline{\mathbf{A}}$ by matrix $\mathbf{A}$ in Eqs. (6.4) and (6.5).

The external dynamic force vector, $\mathbf{f}_o$, can be in different forms resulting from earthquake loading, wind loading, or impulse loading. In the case of the earthquake loading, $\mathbf{f}_o$, is represented by $-\mathbf{M}\ddot{\mathbf{u}}_g$, where $\mathbf{M}$ is the structure mass matrix, $\ddot{\mathbf{u}}_g = \ddot{u}_g \mathbf{I}$, and $\ddot{u}_g$ is the ground acceleration recorded on an accelerograph (accelerogram). In the case of impulse loading, $\mathbf{f}_o$ is in the form $\mathbf{P}\delta(t)$, where $\mathbf{P}$ is the vector of magnitudes of impulsive forces applied at predetermined displacement degrees of freedom of the structure and $\delta(t)$ is the delta function. Wind loading on a structure can be represented as a series of impulse loadings.

Employing the vectorization and microtasking capability of the shared-memory supercomputers such as the Cray YMP8E/8128 (Chapter 2), we have developed an efficient parallel-vector algorithm for the recursive computation of the response of the open-loop and closed-loop systems which is presented in Table 6.1.

6.3 SCHEMES FOR PLACEMENT OF CONTROLLERS

Where should the controllers (actuators and sensors) be placed for optimum control of response of a given structure? This is a significant question specially for large structures with hundreds of members. Only a few researchers have attempted to answer this question. Furuya and Haftka (1995) report that the number of possible combinations for placement of controllers can be very large even for small structures. Stubbs and Park (1996) propose simple rules for the placement of sensors (accelerometers) to regenerate mode shapes of simply supported single-span and continuous two-span beams using the Shannon's sampling theorem (Marks, 1993). Their presentation, however, does not extend to the placement of actuators to control the response of the structure.

Large bridge structures normally have significant stiffness in their longitudinal direction and consequently have substantial in-herent resistance to carry dynamic loadings, say, from earthquakes or winds, in the longitudinal direction without failure (similar to the inherent stiffness of multistory building structures in the vertical direction). In the transverse as well as the vertical directions, however, bridges are quite vulnerable to dynamic loadings. Consequently, we investigate schemes for placement of controllers in order to minimize the response of the bridge in the transverse and vertical directions or in the plane of the bridge cross-section. In the

Table 6.1 Parallel-vector algorithm for the response of the closed loop system.

(The same algorithm can be used for the response of the open loop system of equations by replacing $\overline{A}$ by A.)

Input: n, N, N_{f_o}, n_p, δT, T, $\overline{A}$, B_o, $X^{(0)}$, $f_o^{(0)}$, e, and V.
  (Solution of the complex eigenvalue problem of the unsymmetric matrix $\overline{A}$ is obtained per Tables 4.1 to 4.3 of Chapter 4)

$m = 0$, $i = 2$
FOR $j = 1$ **UNTIL** $2N$ **DO**   (**Vectorization**)

$$X_j^{(m)} = X_j^{(0)}$$

NEXT j
FOR $p = 1$ **UNTIL** n_p **DO**   (**Microtasking**)
  $j = p$
(a)   **FOR** $k = 1$ **UNTIL** $2N$ **DO**   (**Vectorization**)

$$\left[\zeta^{(i)} \right]_{j,k} = I_{j,k}$$

  NEXT k
  $j = j + n_p$
  IF $j \leq 2N$ **THEN** Go to (a)
NEXT p
(b)   **FOR** $p = 1$ **UNTIL** n_p **DO**   (**Microtasking**)

Table 6.1 - continued

$j = p$

(c)   **FOR** $k = 1$ **UNTIL** $2N$ **DO**

$$\left[\zeta^{(i)}\right]_{j,k} = \mathbf{I}_{j,k} + \frac{\delta T \Lambda_{j,l}}{n+1-i} \left[\zeta^{(i-1)}\right]_{l,k}, \, l = 1,2, \, ..., \, 2N$$

(**Vectorization**)

NEXT k

$j = j + n_p$

IF $j \leq 2N$ **THEN** Go to (a)

NEXT p

$i = i + 1$

IF $i \leq n$-1 **THEN** Go to (b)

FOR $p = 1$ **UNTIL** n_p **DO**   (**Microtasking**)

$j = p$

(d)   **FOR** $k = 1$ **UNTIL** $2N$ **DO**

$$\Theta_{j,k} = \mathbf{I}_{j,k} + \delta T \Lambda_{j,l} \left[\zeta^{(n-1)}\right]_{l,k}, \quad l = 1,2, \, ..., \, 2N$$

(**Vectorization**)

NEXT k

FOR $k = 1$ **UNTIL** $2N$ **DO**   (**Vectorization**)

$$\Delta_{j,k} = \int_0^{\delta T} \Theta_{j,k} \, dt$$

Table 6.1 - continued

NEXT k

$j = j + n_p$

IF $j \leq 2N$ **THEN** Go to (d)

NEXT p

$\Gamma = \mathbf{e}\, \Delta\, \mathbf{V}^{T} \mathbf{B}_{\circ}$ **(Microtasking and Vectorization)**

$\Omega = \mathbf{e}\, \Theta\, \mathbf{V}^{T}$ **(Microtasking and Vectorization)**

(e)   **FOR** $p = 1$ **UNTIL** n_p **DO**   **(Microtasking)**

      $j = p$

(f)   **FOR** $k = 1$ **UNTIL** $2N$ **DO**   **(Vectorization)**

$$\mathbf{X}_j^{(m+1)} = \Omega_{j,k} \mathbf{X}_j^{(m)}$$

NEXT k

FOR $k = 1$ **UNTIL** $N_{f_{\circ}}$ **DO**   **(Vectorization)**

$$\mathbf{X}_j^{(m+1)} = \mathbf{X}_j^{(m+1)} + \Gamma_{j,k}\, \mathbf{f}_{\circ}{}_{k}^{(m)}$$

NEXT k

$j = j + n_p$

IF $j \leq 2N$ **THEN** Go to (f)

NEXT p

IF $(m + 1 \leq T/\delta T)$ **THEN**

      FOR $j = 1$ **UNTIL** $2N$ **DO**   **(Vectorization)**

$$\mathbf{X}_j^{(m)} = \mathbf{X}_j^{(m+1)}$$

Table 6.1 - continued

NEXT *j*
$m = m + 1$, Go to (e)
STOP

unlikely case when vibrations in the longitudinal directions need to be controlled as well the required control forces will be minimal and can be handled by installing just two controllers, one at each support.

The dynamic response of a bridge structure in the transverse and vertical directions is mostly due to the first few modes of vibration (usually the first two or three modes of vibrations) (Chopra, 1995). In order to control the vibrations of the bridge in the plane of the bridge cross-section we propose to place four controllers in the plane of the cross-section and investigate three different schemes. The findings of this research are applicable to any kind of bridges consisting of two identical planar structures connected to each other at the top and bottom. But, we limit our examples to three kinds of steel truss bridges: single-span (Figure 6.1), two-span continuous (Figure 6.2), and curved bridges (Figure 6.3). Each bridge consists of two steel trusses, one on each side of the bridge. In a plane of the cross-section, two controllers are placed vertically along the vertical member of each truss and two controllers are placed horizontally along the horizontal members connecting the two trusses at the top and bottom. There are also cross bracings in both lower and upper horizontal planes of the structure in the case of straight bridges (Figures 6.1 and 6.2) and on the top of the structure in the case of the curved bridge (Figure 6.3).

In all the three schemes, the same numbers of actuators and sensors are collocated along the actively-controlled members in the vertical cross-sectional planes of the bridge. In scheme A, four controllers are placed in every vertical planes of the truss passing through the joints except the end planes where only one controller is placed horizontally along the member connecting the two trusses

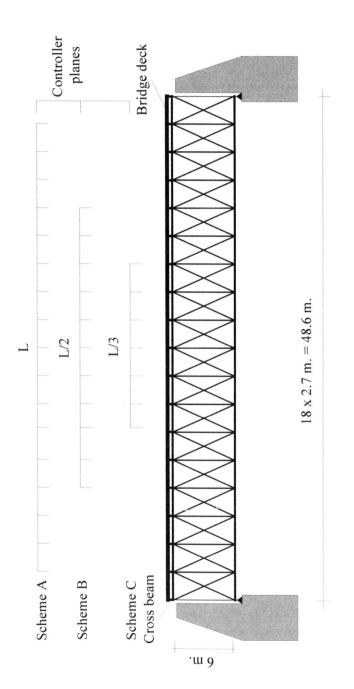

Figure 6.1: Example 1 - One-span truss bridge

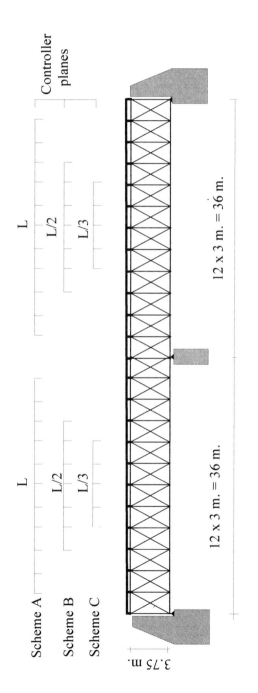

Figure 6.2: Example 2 - Two-span truss bridge

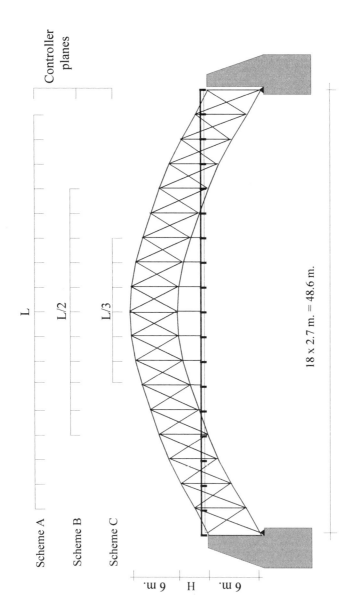

Figure 6.3: Example 3 - One-span curved truss bridge

at the top.

In scheme B, controllers are placed in the vertical planes passing through the joints over the middle half of each span of the bridge when the number of panels in each span is a multiple of four. Otherwise, the middle half of each span is extended from each end up to the adjacent vertical planes of members passing through the joints (for example, the bridge in Figure 6.1 has eighteen 2.7-m panels and we include controllers in the 11 vertical planes of 10 panels). This arrangement is investigated because it covers the locations of the maximum responses of the first two modes of vibrations.

In scheme C, controllers are placed in the vertical planes passing through the joints over the middle third of each span of the bridge when the number of panels in each span is a multiple of six. Otherwise, the middle third of each span is extended from each end up to the adjacent vertical planes of members. This arrangement does not cover the locations of the maximum response of the second mode of vibrations. Investigation of this arrangement will shed light on the significance of the participation of the second mode in the controlled structure.

6.4 EXAMPLES

Three different example bridge structures are investigated. In all the examples, it is assumed that the bridge deck consists of a 7-in. (17.78-cm) concrete deck. Each bridge is initially designed for AASHTO live load of H20 (AASHTO, 1993) and according to the American Institute of Steel Construction (AISC) Allowable Stress Design (ASD) specifications (AISC, 1989). Wide-flange shapes are selected for all the members of the bridge structure using A36 steel with yield stress of 36 ksi (248.2 MPa).

Three kinds of dynamic loadings are considered:

a) A moving vertical impulse loading on each lane of the bridge moving in the same direction with a speed of 65 mph (105 km/h). The magnitude of this load is the resultant of the two axle loads of the AASHTO H20 loading. This load is multiplied by a δ function with a duration of 1 second.

b) The 1940 El Centro earthquake ground acceleration record (Figure 6.4).

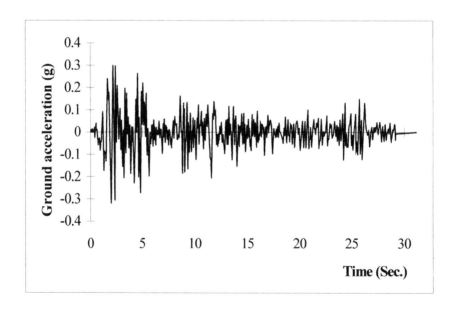

Figure 6.4: **The 1940 El-Centro (California) earthquake ground accelerations.**

c) Periodic impulsive horizontal loadings on each joint of the truss (perpendicular to the plane of the truss) modeling the wind loading on the structure (pressure on one side and suction on the other side), as shown Figure 6.5. The same magnitude is used for all the loads. The magnitude of the wind pressure is found using the AASHTO (1993) code and assuming a wind velocity of 100 miles per hour (160 km/h): $q = 3.59$ kN/m^2. The bridge structure is redesigned for one of the following load combinations satisfying all the AISC ASD and AASHTO stress requirements: a+b or a+c). For wind loading, the AASHTO displacement constraint is also satisfied. The maximum vertical displacement due to the traffic loading and the maximum horizontal displacement due to the wind loading are limited to $L/1000$ where L is the span length. The resulting structure is the open-loop system.

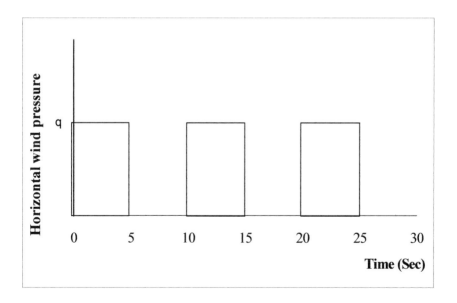

Figure 6.5: Periodic impulsive horizontal wind pressure

Next, controllers are added using one of the three schemes A, B, and C. The response of the closed-loop system is then compared with the response of the open-loop system for the aforementioned dynamic loading cases. We also did consider combination of traffic loading with wind and earthquake loading for the controlled structure (a+b or a+c). Since the traffic loading is in the vertical direction and wind and earthquake loadings are primarily in the horizontal direction we found negligible interactions between the responses of the controlled structure in the horizontal and vertical directions.   Consequently, the effect of load combination is considered insignificant for placement of controllers in bridge structures.

In all the examples, the state variables weighting factor q is chosen equal to one. That is, all state variables are given equal weights. But, different values are chosen for the control weighting factor r in order to investigate the effect of changing the level of control forces on the response of the closed loop structure. Note that there is an inverse relationship between this factor and the level of the force exerted by the actuator as noted in Eq. (3.14).

6.4.1 Example 1

This is a single-span truss bridge with a span of 48.6 m, height of 6 m, and width of 6 m (Figure 6.1). The structure has 292 members and 76 nodes. This results in 432 state variables (3 displacements and 3 velocities for each node of the structure). The three schemes A, B, and C for placement of the controllers are identified in Figure 6.1. Three different values are chosen for the control weighting factor r, starting with $r = 1.0$ and then decreasing it to 0.5 and 0.1.

6.4.2 Example 2

This is a two-span truss bridge with a total span length of 72 m, height of 6 m, and width of 6 m (Figure 6.2). The structure has 388 members and 100 nodes. This results in 564   state variables (3 displacements and 3 velocities for each node of the structure). The three schemes A, B, and C for placement of the controllers are identified in Figure 6.2. The control weighting factor r is chosen equal to 0.1.

6.4.3 Example 3

This is a single-span curved truss bridge with a span of 48.6 m, height of 6 m, and width of 6 m (Figure 6.3) (the same as Example 1). Two different values are chosen for the height H in Figure 6.3: 2 m and 6 m. The structure has 292 members and 76 nodes. This results in 432 state variables (3 displacements and 3 velocities for each node of the structure). The three schemes A, B, and C for placement of the controllers are identified in Figure 6.3. The control weighting factor r is chosen equal to 0.1.

6.5 NUMERICAL RESULTS

In this section, numerical results are presented for three example bridges described in the previous section using three different schemes for actuator placements A, B, and C.

Figure 6.6 displays the vertical displacement at the midpoint of the top chord of the truss of example 1 under the traffic loading

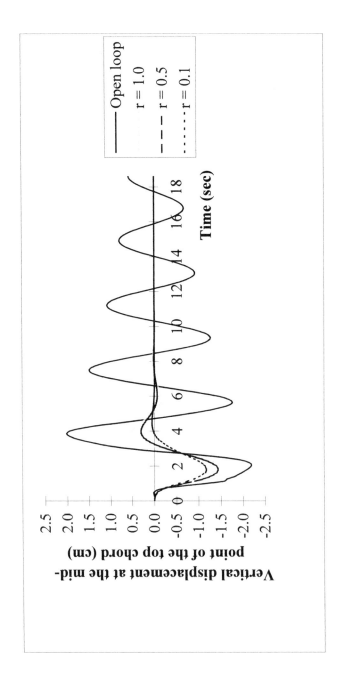

Figure 6.6: The vertical displacement at the midpoint of the top chord of the truss of example 1 under the traffic loading a) for both the open-loop system (uncontrolled structure) and the closed-loop system (controlled structure) employing scheme A and using three different values for the control weighting factor.

a) employing scheme A for both the open-loop system (uncontrolled structure) and the closed loop system (controlled structure) using three different values for the control weighting factor, r. The maximum displacement in the controlled structure is reduced by 30%, 37%, and 50%, for r values of 1.0, 0.5, and 0.1, respectively. It is observed that with a proper control weighting factor it is possible to reduce the maximum response of the controlled structure under traffic loading to 50% of the maximum response of the uncontrolled structure.

Table 6.2 shows the relation between the control weighting factor coefficient and the maximum force exerted by controllers for example 1 under the traffic loading a) employing scheme A. In the rest of this chapter a value of $r = 0.1$ is used for the control weighting factor.

Table 6.2: Relation between the control weighting factor coefficient and the maximum force exerted by the controllers

r	Maximum force (kN)
1.0	1.29
0.5	1.92
0.1	3.81

Using the three different schemes A, B, and C for placement of the controllers, the vertical displacement at the mid-point of the top chord of the truss of example 1 due to traffic loading a) is presented in Figure 6.7. Figures 6.8 and 6.9 present similar results for the horizontal displacement at the same point due to wind loading b) and earthquake loading c), respectively. It is observed that schemes

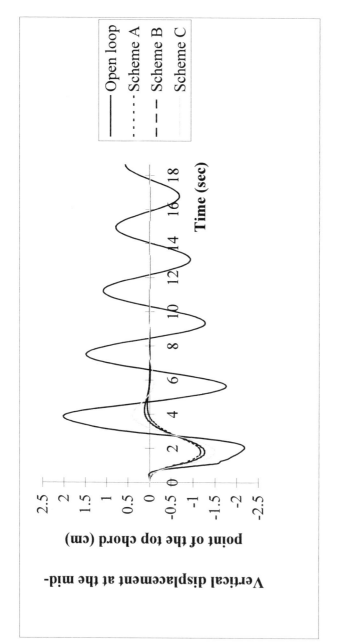

Figure 6.7: The vertical displacement at the midpoint of the top chord of the truss of example 1 due to traffic loading a) using the three different schemes A, B, and C for placement of the controllers.

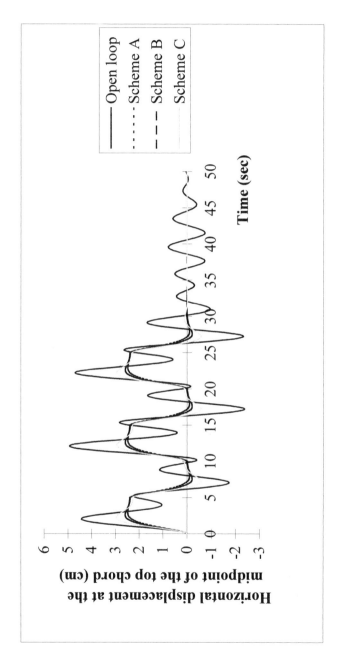

Figure 6.8: The horizontal displacement at the midpoint of the top chord of the truss of example 1 due to wind loading b) using the three different schemes A, B, and C for placement of the controllers.

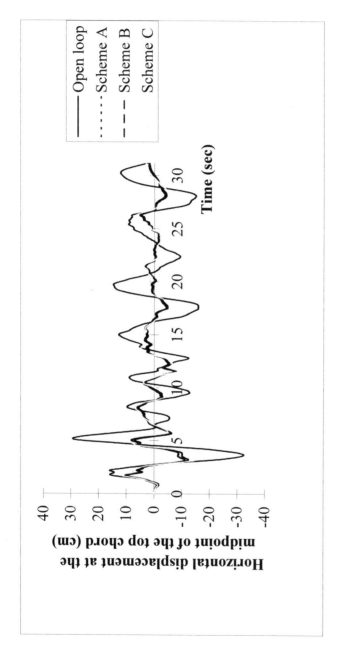

Figure 6.9: **The horizontal displacement at the midpoint of the top chord of the truss of example 1 due to earthquake loading c) using the three different schemes A, B, and C for placement of the controllers.**

A and B reduce the response of the structure more effectively than scheme C, as expected. And scheme A reduces the response only slightly more than scheme B. For the case of wind loading, the maximum response in the controlled structure is reduced by 52%, 48%, and 40% using schemes A, B, and C, respectively. For the case of earthquake loading, the maximum response in the controlled structure is reduced by 73%, 68%, and 55% using schemes A, B, and C, respectively.

Figure 6.10 shows the vertical displacement at the midpoint of the top chord of each span for the two-span bridge of example 2 due to traffic loading a). Figure 6.11 shows the horizontal displacement at the mid-point of the top chord of the left span for example 2 due to wind loading b). The horizontal wind loading is applied to one half of the bridge only in order to obtain the most critical response. Again, it is observed that schemes A and B reduce the response of the structure more effectively than scheme C. And scheme A reduces the response only slightly more than scheme B. The maximum response is reduced by 54%, 50%, and 42% using schemes A, B, and C, respectively. Thus, another conclusion is that controllers can reduce the response of a continuous multi-span bridge more effectively than that of a single-span bridge.

Figure 6.12 displays the vertical displacement at the midpoint of the top chord of the curved truss of example 3 under the traffic loading a) for the open loop system (uncontrolled structure) for three different values of the height H shown in Figure 6.3: 0, 2, and 6m. It is interesting to note that as the height H increases the need for control forces in the vertical direction is reduced. Figure 6.13 displays the horizontal displacement at the mid-point of the top chord of the curved truss of example 3 under the wind loading b) for the

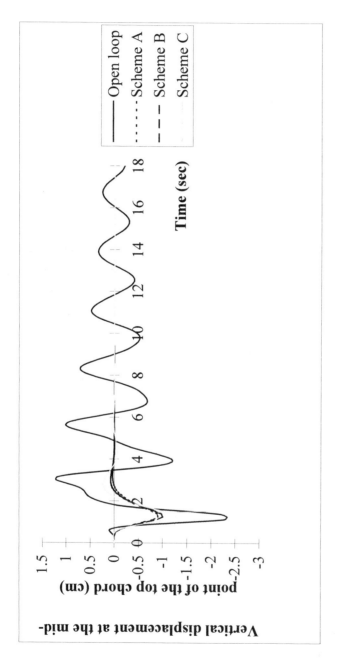

Figure 6.10: The vertical displacement at the midpoint of the top chord of each span of the two-span bridge of example 2 due to traffic loading a)

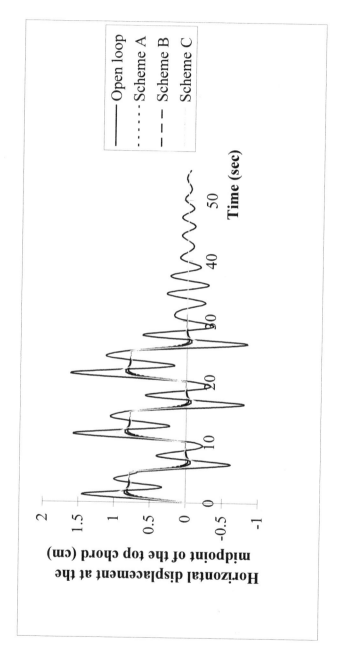

Figure 6.11: The horizontal displacement at the midpoint of the top chord of each span of the two-span bridge of example 2 due to wind loading b)

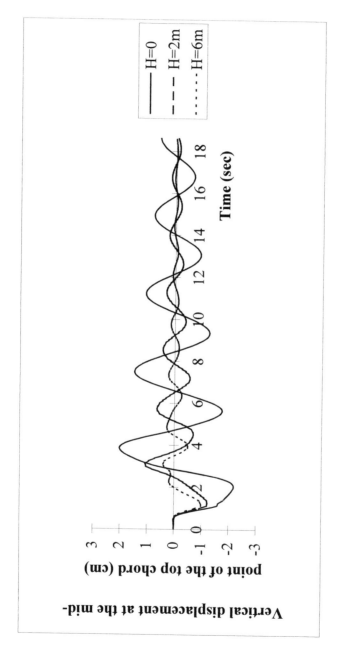

Figure 6.12: The vertical displacement at the midpoint of the top chord of the curved truss of example 3 under the traffic loading a) for the open-loop system for three different values of H shown in Figure 6.3.

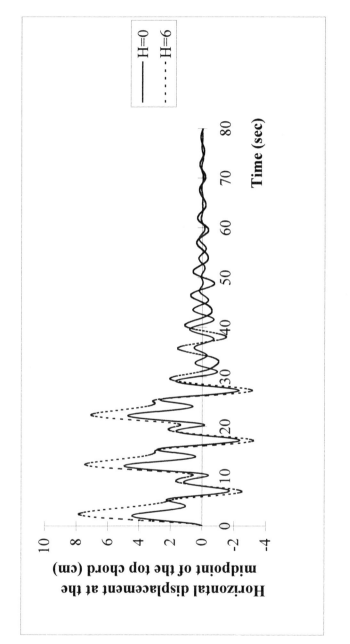

Figure 6.13: The horizontal displacement at the midpoint of the top chord of the curved truss of example 3 due to wind loading b) for the open-loop system for two different values of H shown in Figure 6.3.

open-loop system (uncontrolled structure) for two different values of the height H: 0 and 6m. It is observed that the horizontal displacement increases with the higher value of H, thus, the greater need for controllers in the horizontal direction. Using H=6m, Figures 6.14 and 6.15 present results for the horizontal displacement at the midpoint of the top chord of the curved truss of example 3 due to wind loading b) and earthquake loading c), respectively. It is observed that schemes A and B reduce the response of the structure more effectively than scheme C. And scheme A reduces the response only slightly more than scheme B. For the case of wind loading, the maximum response is reduced by 46%, 42%, and 32% using schemes A, B, and C, respectively. For the case of earthquake loading, the maximum response is reduced by 67%, 62%, and 49% using schemes A, B, and C, respectively.

6.6 CONCLUSIONS

A computational model and high-performance parallel algorithms have been developed for optimal active control of large structures subjected to dynamic loading. The computational model has been applied to three different types of bridges subjected to impulsive traffic, impulsive periodic wind, and dynamic earthquake loadings. Three different schemes have been investigated for placement of controllers. It is demonstrated that through the use of controllers the response of a bridge structure to dynamic loadings can be reduced substantially. The magnitude of the reduction is the largest for the case of dynamic earthquake loading.

On the question of optimal placement of controllers the following conclusions are reached and recommendations made:
- For bridge structures, controllers are recommended to be collocated along the members lying in the cross-sectional

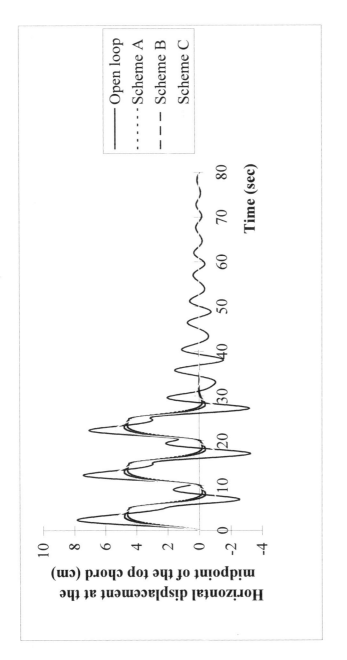

Figure 6.14: The horizontal displacement at the midpoint of the top chord of the curved truss of example 3 due to wind loading b) using the three different schemes A, B, and C for placement of the controllers.

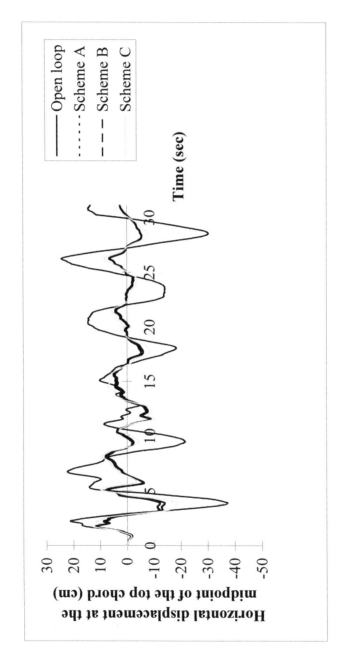

Figure 6.15: The horizontal displacement at the midpoint of the top chord of the curved truss of example 3 due to earthquake loading c) using the three different schemes A, B, and C for placement of the controllers.

planes and perpendicular to the vertical plane of the bridge structure which includes the principal modes of vibrations.

- For different types of bridge structures, schemes A and B reduce the response of the bridge structure more effectively than scheme C. And scheme A is only slightly more effective than scheme B. Since the number of controllers in scheme B is roughly one half of that in scheme A, scheme B is recommended as the more economical solution for placement of controllers.

- Compared with simply supported bridges, the response of continuous bridge structures can be reduced more effectively by controllers. In other words, the required level of actuator forces for continuous bridges is less than that for simply supported bridges.

- The curvature of a curved bridge reduces the response due to the traffic loading in the vertical direction (plane of the curvature), thus reducing the need for controllers to reduce the response in the vertical direction. However, the curvature increases the level of actuator forces needed to reduce the response due to wind and earthquake loadings in the horizontal direction.

CHAPTER 7

SMART MULTISTORY BUILDING STRUCTURES UNDER EARTHQUAKE AND WIND LOADINGS

7.1 INTRODUCTION

In Chapters 4 to 6, we presented a computational model for active control of large adaptive structures subjected to any kind of dynamic loadings. The governing differential equations of the open loop (without controllers) and closed loop (with controllers) systems were formulated and a recursive approach was presented for computing the response of the structure. A robust parallel-vector algorithm was developed for the recursive solution of the response of the open loop and closed loop systems in Chapter 6.

In this chapter, we present the application of the computational model and high-performance parallel algorithms to multistory building structures.

7.2 SCHEMES FOR PLACEMENT OF CONTROLLERS

Placement of controllers for optimum control of response of a structure is of paramount importance. Multistory building structures have significant stiffness in the vertical direction and consequently have substantial inheritant resistance to carry dynamic loadings (for example from earthquakes or wind) in that direction. In the horizontal direction, however, multistory buildings are quite vulnerable to dynamic loadings. Thus, we investigate four different schemes for placement of controllers with the objective of minimizing the response of the building in the horizontal directions.

The dynamic response of a multistory building structure in the horizontal direction is mostly due to the contribution of the first three to six modes of vibrations (Chopra, 1995). In order to control the vibrations of the building structure in the horizontal directions we propose to place the controllers in the horizontal plane of each

floor diaphragm alongside of the beams in two perpendicular
directions (principal axes of the floor plan when there are two axes
of symmetry in the plan). We shall investigate the optimal
distribution of the controllers over the height of the structure.

In all the four schemes, the same numbers of actuators and
sensors are collocated along a preselected number of actively
controlled beams in the horizontal floor planes of the building
structure. In scheme A, controllers are placed alongside of the
beams in every floor plane of the structure. In scheme B,
controllers are placed in every floor plane in the upper half of the
structure. When the number of floors, N, is odd controllers are
placed in the upper $(N+1)/2$ floors of the structure. In scheme C,
controllers are placed in the upper third floors of the structure
when the number of floors is a multiple of three. Otherwise,
controllers are placed in the upper third of the building height
extended one floor down. Finally, in scheme D controllers are
placed in the upper fourth floors of the structure when the number
of floors is a multiple of four. Otherwise, controllers are placed in
the upper fourth of the building height extended one floor down.

7.3 EXAMPLES

Three different example multistory building structures are investigated. The first two examples are space moment-resisting
frames (Figures 7.1 and 7.2). The third example is a space mo-
ment-resistant frame with or without bracings (Figure 7.3). The
actively controlled beams are identified with light grey lines in Figures 7.1 to 7.3. The loading on each structure consists of uniformly distributed dead and live loads of 2.88 kPa (60 psf) and
2.38 kPa (50 psf) (office building), respectively. Each structure is

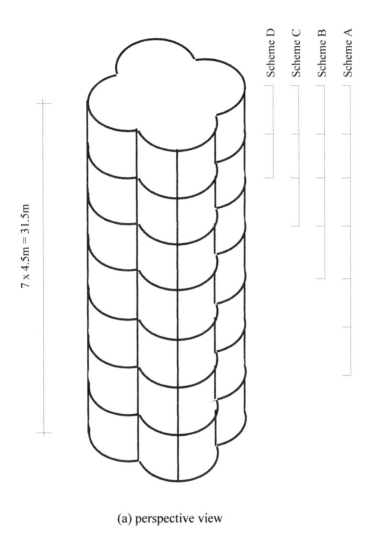

(a) perspective view

Figure 7.1: Example 1 Seven-story space moment-resisting frame with a cloverleaf plan

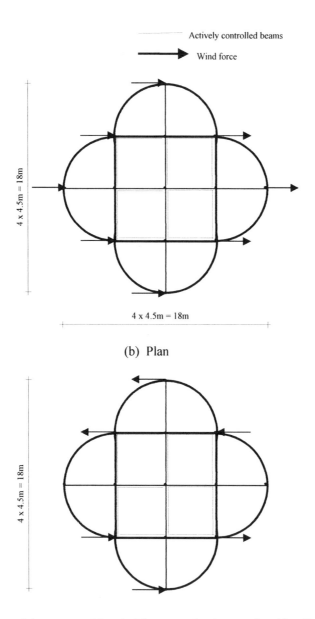

(b) Plan

(c) unsymmetric wind forces to simulate torsional loading

Figure 7.1 - continued

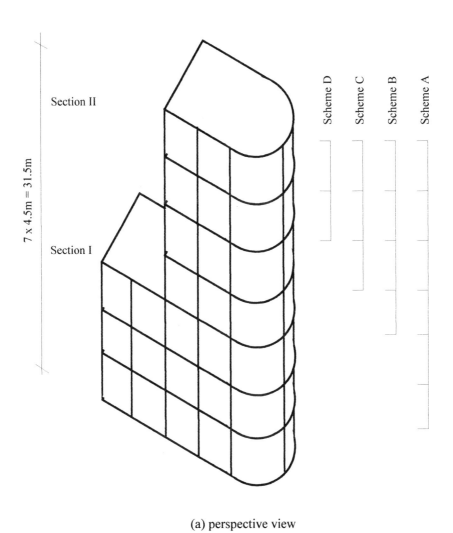

(a) perspective view

**Figure 7.2: Example 2 - Seven-story space moment-resisting
frame with setback**

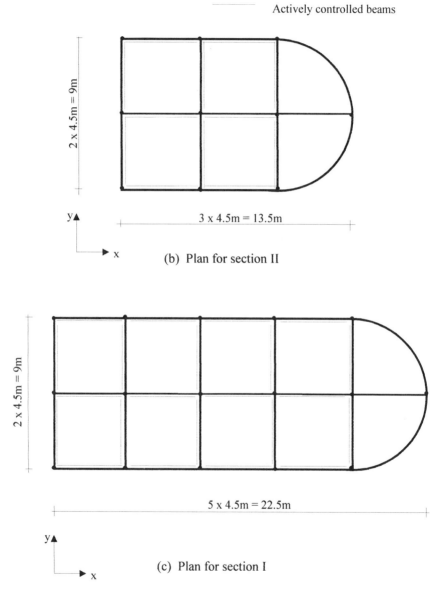

Actively controlled beams

(b) Plan for section II

(c) Plan for section I

Figure 7.2 - continued

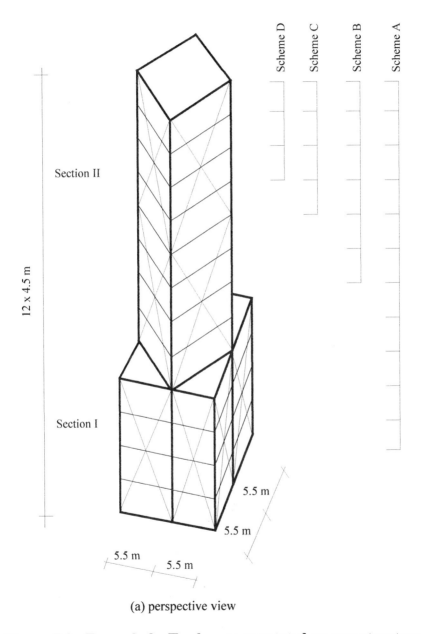

(a) perspective view

Figure 7.3: Example 3 - Twelve-story rotated-square structure

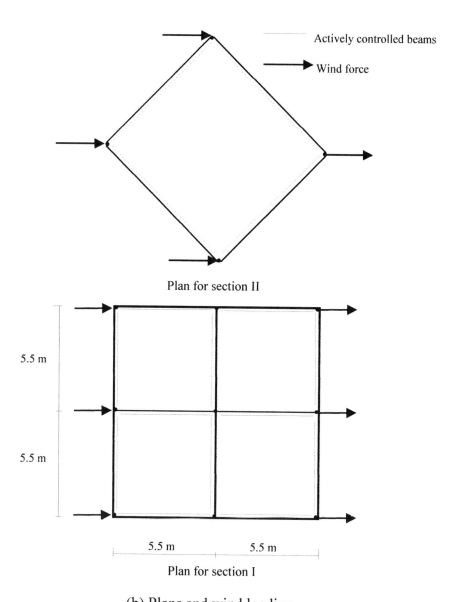

Plan for section II

Plan for section I

(b) Plans and wind loading

Figure 7.3 - continued

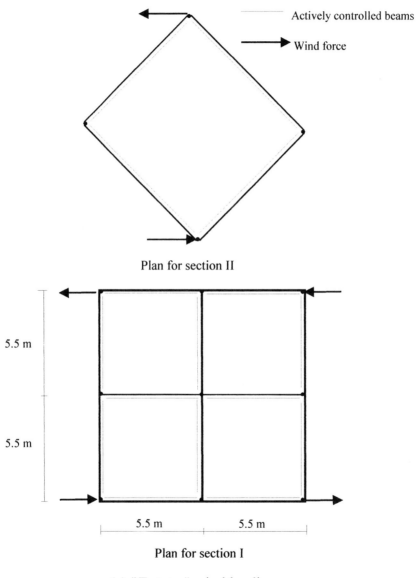

Actively controlled beams

Wind force

Plan for section II

5.5 m

5.5 m

5.5 m   5.5 m

Plan for section I

(c) "*Twister*" wind loading

Figure 7.3 - continued

initially designed for dead plus live loads according to the AISC
ASD specifications (AISC, 1989). Wide-flange shapes from the
AISC manual (AISC, 1989) are selected for all the members
including bracings.

Curved members in examples 1 and 2 are designed for
combined biaxial bending and torsion. The loading on each curved
beam is approximately parabolic (Figure 7.4a). An approximate
method is used for design for torsion. Consider the equilibrium of a
differential element of a curved beam with length ds (Figure 7.4b).
By taking moments about a tangential plane, we obtain
(Brockenbrough and Merritt, 1994)

$$\frac{dT}{ds} = \frac{M}{R} \qquad\qquad (7.1)$$

where M is the bending moment due to the vertical load (Figure
7.4a), T is the torque, and R is the radius of curvature. Thus, we
first compute the bending moment (Figure 7.4d) due to the vertical
loads. Then, all ordinates are divided by the radius R. Next, the
resulting M/R diagram is integrated over the curved length of the
beam to find the torques T (Figure 7.4e). The torque produces
warping normal stresses (bending with respect to the minor axis of
the cross-section) in the flanges. These stresses are added to the
direct bending stresses. Further, the torque produces additional
shear force and stresses defined by the following equation:

$$V = \frac{dM}{ds} + \frac{T}{R}. \qquad\qquad (7.2)$$

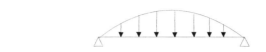

(a) Vertical loading of a curved beam

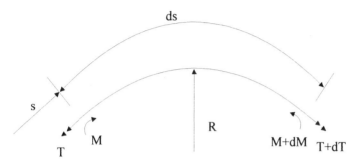

(b) Plan of a differential element of a curved beam

(c) Elevation of a differential element of a curved beam

(d) Bending moment diagram

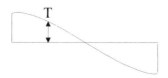

(e) Torsional moment diagram

Figure 7.4: Bending and torsion of a curved beam

These shear stresses are also added to the shear stresses resulting from the bending.

Three kinds of dynamic loadings are considered:

a) The 1940 El Centro (California) earthquake ground motion (Figure 6.7).

b) Periodic impulsive horizontal loadings (Figure 6.8) on each exterior joint of the structure (pressure on one side and suction on the other side) modeling the wind loading on the structure (see Figure 7.1b, for Example 1). The magnitude of the wind pressure is computed according to the Uniform Building Code (UBC, 1994) based on a basic wind speed of 113 km/h (70 mph), exposure C (generally open area), and an importance factor of 1.

c) Similar to b) but wind forces are applied in one direction on one half of the structure (say, west-east direction), and on the opposite direction (say, east-west direction) on the other half of the structure (Figure 7.1c) in order to simulate approximately unsymmetric torsional loading or a *twister*.

The building structure is redesigned for a combination of static dead and live loads plus one of the aforementioned dynamic loadings. The resulting structure is the open-loop system.

Next, controllers are added using one of the four schemes A, B, C, and D described earlier. The response of the closed-loop system is then compared with the response of the open-loop system for the aforementioned dynamic loading cases.

Different values are chosen for the control weighting factor r (see Section 6.2) in order to investigate the effect of changing the level of control forces on the response of the closed loop structure for different types of dynamic loadings. There is an inverse relationship between the weight factor and the required level of actuator forces.

The displacement response curves presented in Figures 7.5 to 7.15 are all the maximum displacement at the top of the structure.

7.3.1 Example 1

This example is a 7-story moment-resisting frame with a cloverleaf plan and a height of 31.5 m , as shown in Figure 7.1. The structure has 259 members and 91 nodes (excluding supports). This results in 1092 state variables (6 displacements and 6 velocities for each node of the structure). The four schemes A, B, C, and D for placement of the controllers are identified in Figure 7.1.

Figure 7.5 displays the horizontal displacement at the top of the structure due to the earthquake loading a) employing scheme A for both the open loop system (uncontrolled structure) and the closed loop system (controlled structure) using three different values for the control weighting factor, r. The maximum displacement in the controlled structure is reduced by 27%, 53%, and 86%, for r values of 0.05, 0.01, and 0.001, respectively. These values correspond to maximum force exerted by controllers of 75.9 kN, 166.1 kN, and 318.9 kN, respectively. It is observed that with a proper selection of the control weighting factor it is possible to reduce the maximum response of the controlled structure under earthquake loading to a fraction of the maximum response of the uncontrolled structure.

Using the four schemes A, B, C, and D for placement of the controllers, the horizontal displacement due to earthquake loading a) using a value of 0.01 for the control weighting factor is presented

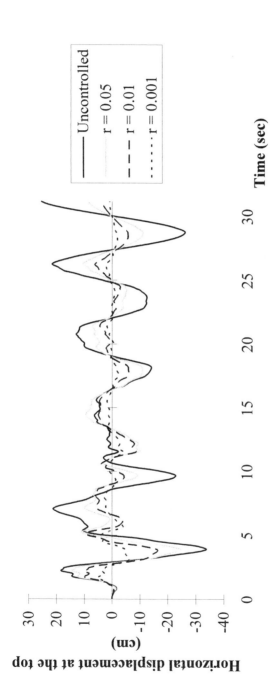

Figure 7.5: The horizontal displacement at the top of the structure of example 1 due to earthquake loading a) for both the open-loop system (uncontrolled structure) and the closed-loop system (controlled structure) employing scheme A and using three different values for the control weighting factor.

in Figure 7.6. It is observed that schemes A B, and C reduce the response of the structure more effectively than scheme D. And schemes A and B reduce the response by roughly the same amount. They reduce the response only slightly more than scheme C. The maximum displacement in the controlled structure is reduced by 53%, 52%, 49%, and 44% using schemes A, B, C, and D, respectively. These values correspond to maximum force exerted by controllers of 166.1 kN, 170.1 kN, 180.7 kN, and 191.9 kN, respectively.

Using the four schemes A, B, C, and D for placement of the controllers, the horizontal displacement due to wind loading b) using a value of 0.00005 for the control weighting factor is presented in Figure 7.7. It is observed that schemes A B, and C reduce the response of the structure more effectively than scheme D. The three schemes A, B, and C reduce the response by roughly the same amount. The maximum displacement in the controlled structure is reduced by 96%, 96%, 95%, and 89% using schemes A, B, C, and D, respectively. These values correspond to maximum force exerted by controllers of 160.3 kN, 236.1 kN, 301.5 kN, and 357.6 kN, respectively. It is also observed that the maximum control force required increases substantially as we move from scheme A to scheme D.

We observed the same conclusions when using the four schemes A, B, C, and D for placement of the controllers using a value of 0.00001 for the control weighting factor with unsymmetric wind loading c). The maximum displacement in the controlled structure is reduced by 80%, 80%, 78%, and 67% using schemes A, B, C, and D, respectively. These values correspond to maximum force exerted by controllers of 192.8 kN, 200.6 kN, 202.2 kN, and 203.6 kN, respectively.

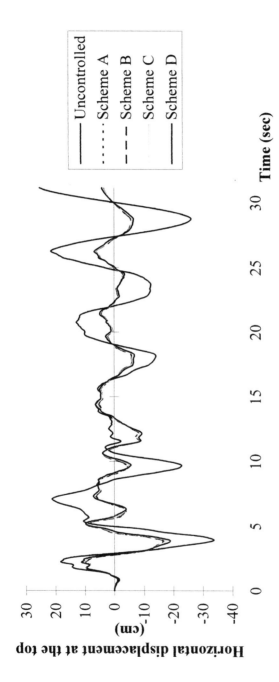

Figure 7.6:   **The horizontal displacement at the top of the structure of example 1 due to earthquake loading a) using the four different schemes A, B, C, and D for placement of controllers**

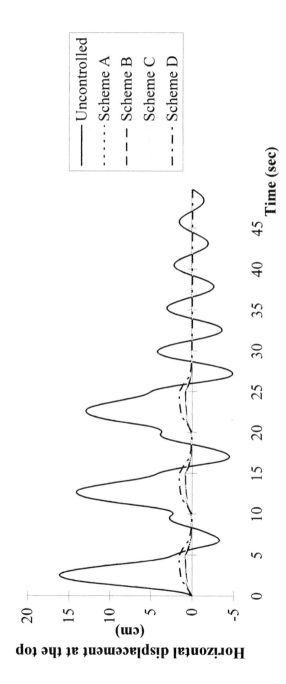

Figure 7.7: The horizontal displacement at the top of the structure of example 1 due to wind loading b) using the four different schemes A, B, C, and D for placement of controllers

7.3.2 Example 2

This example is a 7-story moment-resisting frame with an irregular plan and setback and a height of 31.5 m, as shown in Figure 7.2. The structure has 223 members and 88 nodes (excluding supports). This results in 1056 state variables (6 displacements and 6 velocities for each node of the structure). The four schemes A, B, C, and D for placement of the controllers are identified in Figure 7.2.

Using the four schemes A, B, C, and D for placement of the controllers, the horizontal displacement due to wind loading b) in the x-direction (Figure 7.2) using a value of 0.00005 for the control weighting factor is presented in Figure 7.8. An interesting observation is made here. That is, for an irregular structure such as example 2, the four schemes produce completely different responses, with the scheme A being the most effective and scheme D being the least effective. The maximum displacement in the controlled structure is reduced by 81%, 68%, 51%, and 33% using schemes A, B, C, and D, respectively. These values correspond to maximum force exerted by controllers of 168.3 kN, 197.1 kN, 224.3 kN, and 226.4 kN, respectively. We observed the same conclusions with the other two cases of dynamic loadings, earthquake loading and unsymmetric wind (twister) loading.

7.3.3 Example 3

This example is a 12-story moment-resisting frame with a rotated-square plan and a height of 54 m , as shown in Figure 7.3. The structure has 152 members and 68 nodes (excluding supports).

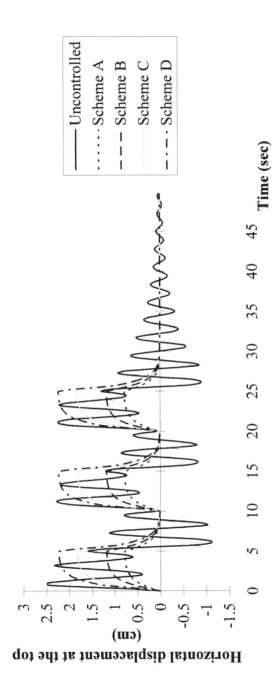

Figure 7.8: The horizontal displacement at the top of the structure of example 2 due to wind loading b) using the four different schemes A, B, C, and D for placement of controllers

This results in 816 state variables (6 displacements and 6 velocities for each node of the structure). The four schemes A, B, C, and D for placement of the controllers are identified in Figure 7.3. In our example 3a no bracing is used. In example 3b, cross bracings are used to study the effect of bracings on the response of the controlled structure (Figure 7.3). Three different arrangements for cross bracings are investigated. In all cases, each bracing is placed on the exterior and covers four stories of the structure (Figure 7.3a). In the first arrangement, I, cross bracings are placed in the lower third of the building height (4 floors). In the second arrangement, II, cross bracings are placed in the lower two thirds of the structure. Finally, in the third arrangement, III, cross bracings are placed over the entire height of the structure.

Using the four schemes A, B, C, and D for placement of the controllers, the maximum horizontal displacements at the top of the structure due to wind loading b) using a value of 0.0005 for the control weighting factor and due to unsymmetric wind loading c) (twister) using a value of 0.00005 for the control weighting factor are presented in Figures 7.9 and 7.10, respectively. It is observed that for a relatively slender structure such as example 3, the four schemes produce completely different responses, with the scheme A being the most effective and scheme D being the least effective. The maximum displacement in the controlled structure due to wind loading b) is reduced by 79%, 75%, 73%, and 71% using schemes A, B, C, and D, respectively. These values correspond to maximum force exerted by controllers of 143.9 kN, 152.9 kN, 157.6 kN, and 164.5 kN, respectively. The maximum displacement in the controlled structure due to unsymmetric wind loading c) (twister) is reduced by 87%, 75%, 61%, and 54% using schemes A, B, C, and D, respectively. These values correspond to maximum force exerted

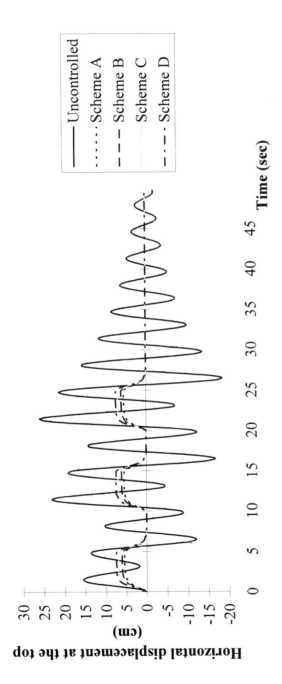

Figure 7.9: The horizontal displacement at the top of the structure of example 3a (without cross bracings) due to wind loading b) using the four different schemes A, B, C, and D for placement of controllers

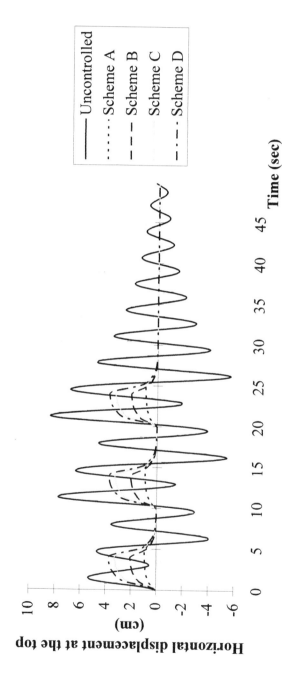

Figure 7.10: The maximum horizontal displacement at the top of the structure of example 3a (without cross bracings) due to unsymmetric wind loading c) (twister) using the four different schemes A, B, C, and D for placement of controllers

by controllers of 138.2 kN, 203.6 kN, 222.9 kN, and 233.3 kN, respectively.

Figure 7.11 displays the horizontal displacement due to the earthquake loading a) employing scheme A for both uncontrolled unbraced and controlled braced structures using the bracing arrangements, I, II, and III. The maximum displacement in the controlled structure is reduced by 47%, 78%, and 85% using arrangements I, II, and III, respectively. It is observed that bracing arrangements II and III in the controlled structure reduce the displacement more effectively than bracing arrangement I. And bracing arrangement III reduces the displacement only slightly more than bracing arrangement II.

Using the four schemes A, B, C, and D for placement of the controllers and bracing arrangement I, the maximum horizontal displacements due to wind loading b) using a value of 0.0005 for the control weighting factor and due to unsymmetric wind loading c) using a value of 0.00005 for the control weighting factor are presented in Figures 7.12 and 7.13, respectively. For the case of the unsymmetric wind loading, it is observed that the four schemes produce completely different responses, with the scheme A being the most effective and scheme D being the least effective (Figure 7.13). However, scheme A reduces the response only slightly more than scheme B. The maximum displacement in the controlled structure due to wind loading b) is reduced by 81%, 79%, 74%, and 73% using schemes A, B, C, and D, respectively. The maximum displacement in the controlled structure due to wind loading c) is reduced by 88%, 81%, 63%, and 55% using schemes A, B, C, and D, respectively. It is also observed that controllers are more effective in reducing the response in unbraced frames than braced frames.

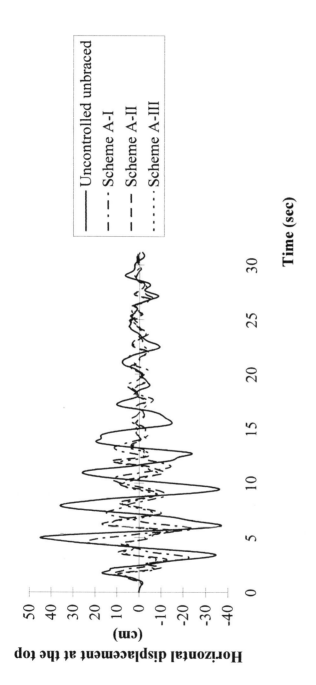

Figure 7.11: The horizontal displacement at the top of the structure of example 3b due to earthquake loading a) using the scheme A for placement of the controllers with the three different bracing arrangements I, II, and III

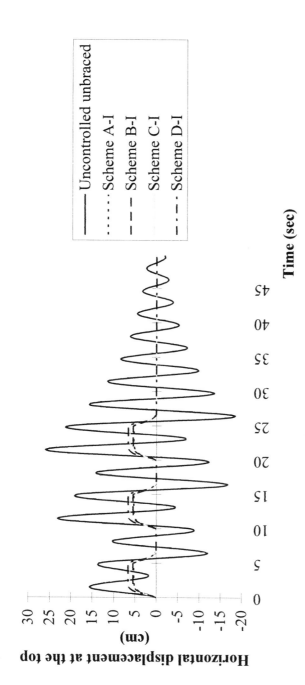

Figure 7.12: The horizontal displacement at the top of the structure of example 3b due to wind loading b) using the four different schemes A, B, C, and D for placement of controllers and bracing arrangement I

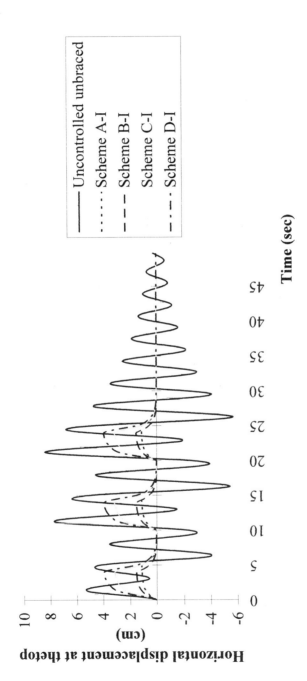

Figure 7.13: The maximum horizontal displacement at the top of the structure of example 3b due to unsymmetric wind loading c) using the four different schemes A, B, C, and D for placement of the controllers and bracing arrangement I

Figure 7.14 displays the horizontal displacement due to the earthquake loading a) for uncontrolled fully braced structure (example 3b) and controlled structure without bracings (example 3a) using the two schemes A and D for placement of the controllers and using a value of 0.008 for the control weighting factor. The maximum displacement in the controlled unbraced structure using scheme A is 2.6% less than that in the uncontrolled braced structure (example 3b). The maximum displacement in the uncontrolled braced structure (example 3b) is 12% less than that in the controlled unbraced structure (example 3a) using scheme D. It is observed that scheme A in the controlled unbraced structure reduces the maximum displacement more than bracings in the braced uncontrolled structure. It was also found that scheme B reduces the response by roughly the same amount as bracings.

7.4 CONCLUSIONS

The computational model and high-performance parallel algorithms developed in this work for optimal control of adaptive structures have been applied to three different multistory steel building structures subjected to dynamic earthquake and symmetric and unsymmetric periodic impulsive wind loadings. They include irregular structures with setbacks and curved beams. Four different schemes have been investigated for placement of controllers. It is demonstrated that through proper selection of the control weighting factor the response of a multistory building structure can be reduced to a fraction of the response of the uncontrolled structure. This can be achieved at the practically feasible maximum required actuator force of around 200 kN.

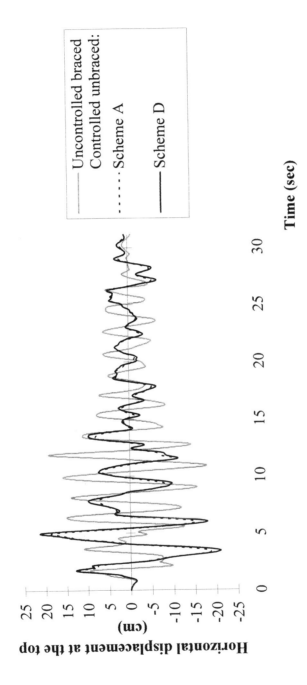

Figure 7.14: The horizontal displacement due to the earthquake loading a) for uncontrolled fully braced structure and controlled unbraced structure using the two schemes A and D for placement of controllers

The following conclusions are reached and recommendations made:

- For multistory building structures, controllers are recommended to be collocated along the beams in the horizontal floor diaphragms in two perpendicular directions (principal axes of the floor plan when there are axes of symmetry in the plan).

- For a regular structure with a low aspect ratio such as example 1, schemes A, B, and C reduce the response of the structure substantially more effectively than scheme D. The effectiveness of schemes A, B, and C in reducing the response is practically the same. Consequently, scheme C is recommended as the optimal scheme because it requires the minimum number of controllers among the three schemes.

- For irregular structures such as example 2 the four different schemes reduce the response differently with scheme A to be the most effective and schemes D to be the least effective. Depending on the level of irregularity, scheme A or B would be recommended.

- For slender structures with a relatively high aspect ratio such as example 3, the four schemes reduce the response differently. The differences, however, are not as great as those for an irregular structure. The optimal arrangement for placement of controllers depends on the height and aspect ratio of the structure. Depending on the height and aspect ratio, scheme A or B would be recommended.

- Controllers are more effective in reducing the response in unbraced frames than braced frames.

- Controllers reduce the need for bracings substantially, thus making the structure architecturally more pleasing.

CHAPTER 8

SMART BUILDING STRUCTURES UNDER BLAST LOADING

8.1 INTRODUCTION

With the recent bombing of the World Trade Center in New York City and the Federal Building in Oklahoma City and other structures the question of designing bomb-resistant structures has resurfaced (Fental, 1996). In other words, an increasing number of structures have to be designed not only to resist dynamic forces emanating from natural sources such as winds and earthquake but also human destructive actions such as blast loading and bombs.

In the previous chapters, the authors presented the mathematical foundations, computational models, and parallel algorithms for creation of a new generation of high-performance adaptive/smart structures with self-modification capability to resist any kinds of dynamic loading. This work is based on ingenious integration of five different disciplines/technologies: structural engineering, control, mathematical optimization, sensor/actuator technology, and high-performance computing.

In this chapter, we apply the computational model and parallel algorithms presented in the previous chapters for optimal control of large multistory building structures subjected to blast and bomb loadings and demonstrate how the structural response can be reduced substantially through the proper use of actively controlled members. Both internal blast loading at different floor levels and external blast loading from outside the structure are considered. Results are presented for several regular and irregular steel moment-resisting space frame structures.

8.2 BLAST LOADING

In order to simulate an actual bomb blast situation, bars of pentaerythritoltetra-nitrate (PETN) are used as explosives (Britton,

1983 and Britton et al, 1984). A continuous impulsive pressure of variable magnitude shown in Figure 8.1 results from such an explosion. The effective blast time is in the order of 1.0 second. In this work, the blast loading is modeled as continuous impulsive forces applied at the joints of the structure.

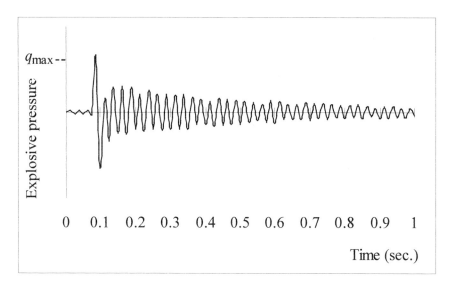

Figure 8.1: Impulsive blast pressure.

Blasts from both within the structure and from outside the structure are considered. For the blast from outside the building, the impulsive forces are applied at every joint on the outside of the structure facing the explosives. During an explosion, different types of energy are released. The type of energy that affects structures the most is the shock energy which is the result of the detonation pressure exerted by the detonation wave propagating through the explosive column (bar). The detonation pressure is a function of the explosive density times the explosive detonation velocity squared (Konya and Walter, 1990). The detonation velocity is the speed at which the detonation moves through the column of explosives. It ranges from approximately 1,524 to 7,620 m (5,000 to 25,000 ft) per second for commercially used products. Determination of the detonation pressure is complex; it can be estimated only approximately through experimentation. An approximate formula is used in this work to determine the detonation pressure (Konya and Walter, 1990)

$$q_n = 4.18 \times 10^{-7} \gamma v^2 /(1 + 0.8\gamma) \tag{8.1}$$

where, q_n = the detonation pressure in kilobars (1 kilobar = 14,504 psi = 99,653 kPa), γ = the specific gravity of the explosive, and v = the detonation velocity in ft per sec (1 ft = 0.3048 m).

The impulsive forces at various joints are computed assuming that their magnitude is reduced exponentially with respect to the distance, z, from the explosives to the joint of the structure,

$$q(z) = q_n \exp(-z) \tag{8.2}$$

and applied in the direction of a vector connecting the center of explosives to the joint. As such, three components of the impulsive force are applied at each joint of the structure, two in the horizontal directions, and one in the vertical direction. Under this scenario the building structure will vibrate in three different directions with the vibrations in the vertical directions to be usually insignificant due to the large stiffness in the vertical direction of the structure. Out of the two horizontal components of vibrations, the one perpendicular to the outside plane of the structure facing the explosives is the dominant one. This three-dimensional dynamic loading and the resulting three-dimensional vibrations are fundamentally different from the customary approach of assuming one-dimensional horizontal loadings due to wind or earthquake loadings. In other words, the assumption of loading only along a single axis of the structure is not valid for blast loading.

For blast within the structure, it is assumed that the explosives are placed on a floor of the structure and the impulsive forces are applied to all the joints in the story including that particular floor.

8.3 PLACEMENT OF CONTROLLERS

As discussed in Chapter 7, multistory building structures have significant stiffness in the vertical direction and consequently have substantial inherent resistance to carry dynamic loadings in that direction. In the horizontal direction, however, multistory buildings are quite vulnerable to dynamic loadings. In order to control the vibrations of the building structure in the horizontal plane we place the controllers in the horizontal plane of each floor diaphragm in two perpendicular directions (principal axes of the floor plan when there are axes of symmetry in the plan). Since the location of the explosives can not be known in advance controllers are placed along the members in the horizontal floor plane of every floor of the structure.

8.4 EXAMPLES

Three different example multistory building structures with different aspect ratio are investigated. The three examples are steel space moment-resisting frames (Figures 8.2, 8.3, and 8.4). The actively controlled beams are identified with dashed lines in the floor plans in Figures 8.2, 8.3, and 8.4. The loading on each structure consists of uniformly distributed dead and live loads of 2.88 kPa (60 psf) and 2.38 kPa (50 psf) (office building), respectively. Each structure is initially designed for dead plus live loads according to the AISC ASD specifications (AISC, 1989). Wide-flange shapes from the AISC manual (AISC, 1989) are selected for all the members.

Curved members in examples 2 and 3 are designed for combined biaxial bending and torsion, as explained in Section 7.3.

8.4.1 Example 1

This example is a 12-story moment resisting frame with a rotated-square plan, as shown in Figure 8.2. Properties of this structure are presented in Section 7.3.3. Four different cases of blast loadings are investigated. In the case one, two, and three, a bomb is exploded on the fourth, eighth, and twelfth floor, respectively in order to study the effect of placing a bomb on different floors on the response of the structure with a maximum pressure of 65 MPa (q_{max} in Figure 8.1). In the fourth case, a bomb is exploded from the outside of the building located at two different distances of 5 m and 2.5 m from the front of the structure.

8.4.2 Example 2

This example is a 7-story steel space moment-resisting frame with a cloverleaf plan, as shown in Figure 8.3. Properties of this structure are presented in Section 7.3.1. A bomb is exploded on the seventh floor with a maximum pressure of 65 MPa.

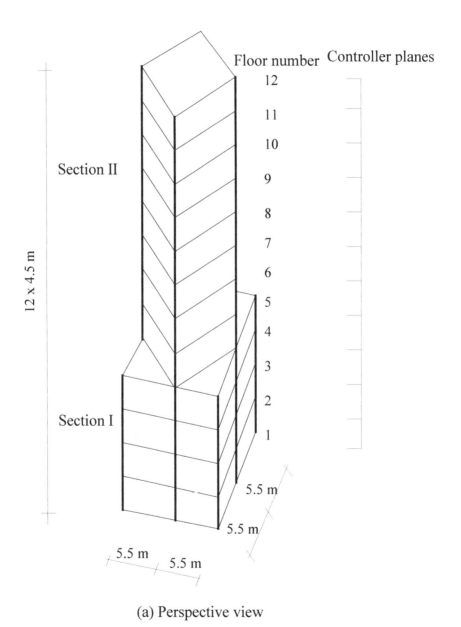

(a) Perspective view

Figure 8.2 Example 1 - Twelve-story rotated-square structure

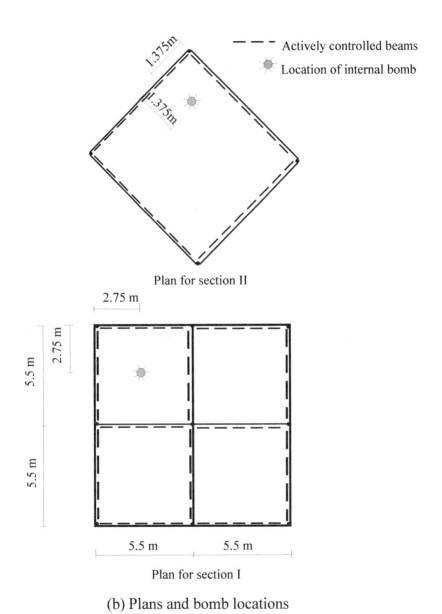

Plan for section II

(b) Plans and bomb locations

Figure 8.2 - continued

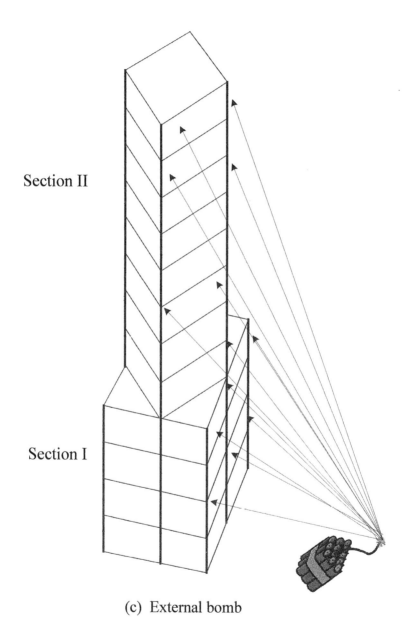

Section II

Section I

(c) External bomb

Figure 8.2 - continued

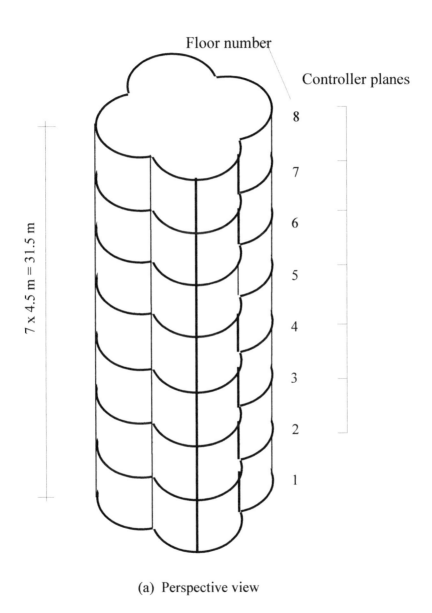

(a) Perspective view

Figure 8.3: Example 2 - Seven-story space moment-resisting frame with a cloverleaf plan

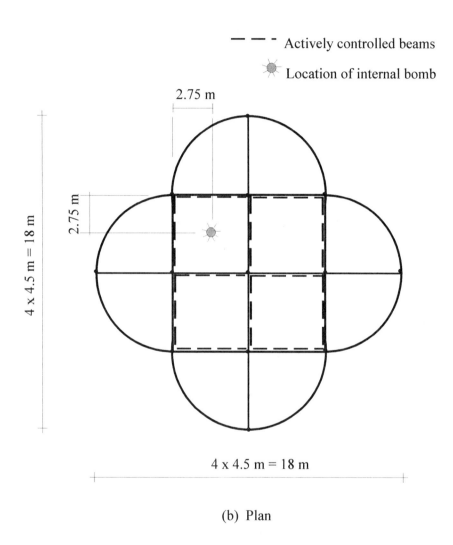

(b) Plan

Figure 8.3 - continued

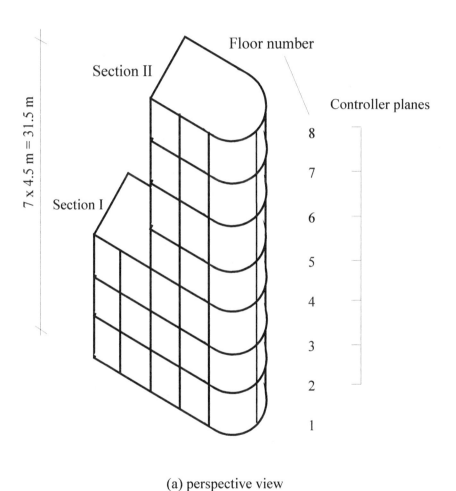

(a) perspective view

Figure 8.4: Example 3 - Seven-story irregular structure

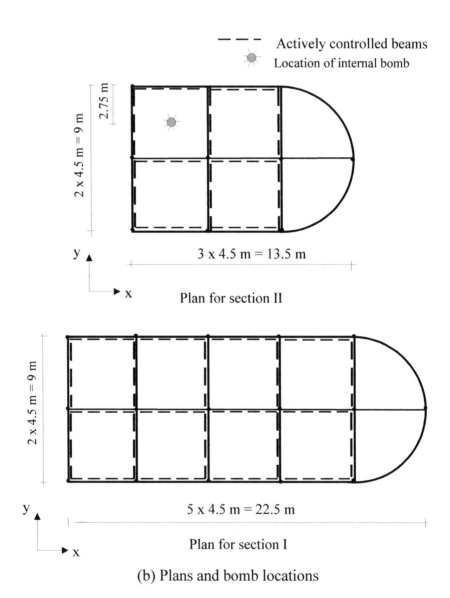

(b) Plans and bomb locations

Figure 8.4 - continued

8.4.3 Example 3

This example is a 7-story moment-resisting frame with an irregular plan and setback, as shown in Figure 8.4. Properties of this structure are presented in Section 7.3.2. A bomb is exploded on the seventh floor with a maximum pressure of 65 MPa.

8.5 RESULTS AND DISCUSSION

8.5.1 Example 1:

Figure 8.5 shows The horizontal displacement at the fourth floor of the structure of example 1 due to the explosion of a bomb on the fourth floor for both the open loop system (uncontrolled structure) and the closed loop system (controlled structure). It is shown that with an appropriate choice of the level of control forces the response due to the internal blast loading can be reduced substantially. In this case, the maximum displacement is reduced by 59%. Figure 8.6 shows the envelope of the maximum lateral displacements of the uncontrolled structure of example 1 due to the explosion of a bomb on the fourth floor. It is observed that when an explosion happens on a floor at the lower third inside the building, the maximum lateral displacements occur at the level of explosion. Figures 8.7 and 8.8 show the lateral displacements of both the uncontrolled and controlled structures of example 1, respectively, due to the explosion of a bomb on the fourth floor at time intervals of 0.0125 sec (curve 1), 0.0375 sec (curve 2), 0.0625 sec (curve 3), 0.0875 sec (curve 4), and 0.1125 sec (curve 5).

Figure 8.9 shows the horizontal displacement at the top of the structure of example 1 due to the explosion of a bomb on the twelfth floor for both the uncontrolled and the controlled

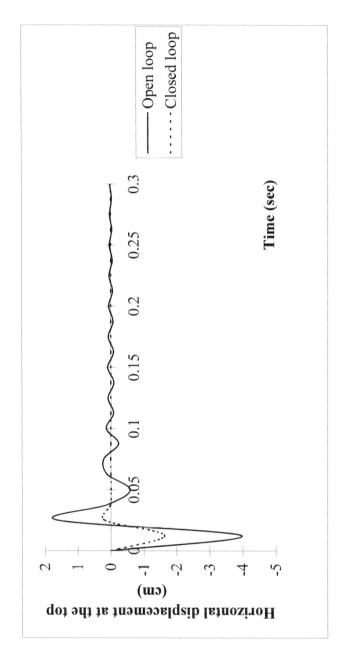

Figure 8.5: The horizontal displacement at the fourth floor of the structure of example 1 due to the explosion of a bomb on the fourth floor for both the open-loop system (uncontrolled structure) and the closed loop system (controlled structure).

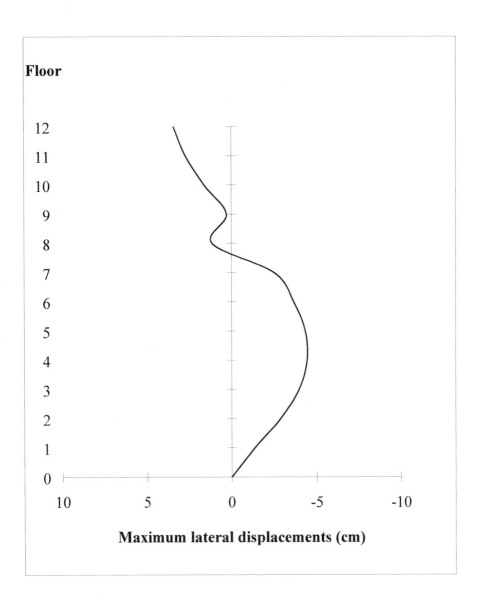

Figure 8.6: Envelope of the maximum lateral displacement of the uncontrolled structure of example 1 due to the explosion of a bomb on the fourth floor.

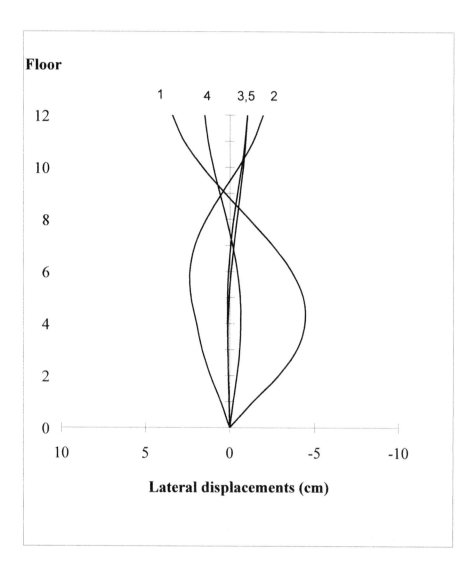

Figure 8.7: Lateral displacements of the uncontrolled structure
            of example 1 due to the explosion of a bomb on the
            fourth floor at time intervals of 0.0125 sec (curve 1),
            0.0375 sec (curve 2), 0.0625 sec (curve 3), 0.0875 sec
            (curve 4), and 0.1125 sec (curve 5).

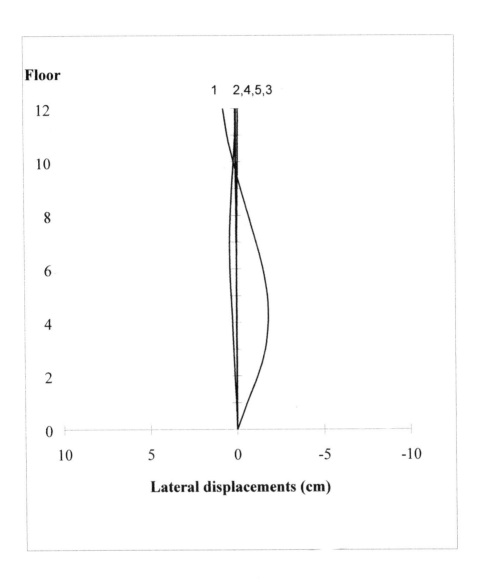

**Figure 8.8: Lateral displacements of the controlled structure of
example 1 due to the explosion of a bomb on the
fourth floor at time intervals of 0.0125 sec (curve 1),
0.0375 sec (curve 2), 0.0625 sec (curve 3), 0.0875 sec
(curve 4), and 0.1125 sec (curve 5).**

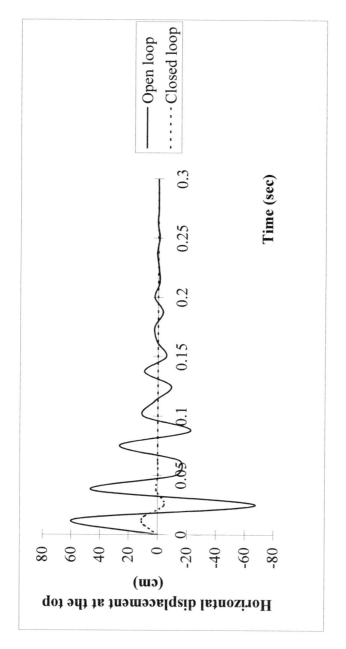

Figure 8.9:The horizontal displacement at the top of the structure of example 1 due to the explosion of a bomb on the twelfth floor for both the uncontrolled and the controlled structures.

structures. It is observed that the maximum lateral displacements occur at the top of the structure. Figures 10 and 11 show the lateral displacements of the both uncontrolled and controlled structures of example 1, respectively, due to the explosion of a bomb on the twelfth floor at time intervals of 0.0125 sec (curve 1), 0.0375 sec (curve 2), 0.0625 sec (curve 3), 0.0875 sec (curve 4), and 0.1125 sec (curve 5).

Figure 8.12 shows the horizontal displacement at the top of the uncontrolled structure of example 1 due to the explosion of a bomb outside the building at distances 2.5 and 5.0m. from the building, and the controlled structure when a bomb is exploded at a distance 2.5 m. from the building. It is observed that as the bomb gets closer to the building its effect increases dramatically.

8.5.2 Example 2:

Figure 8.13 shows the horizontal displacement at the top of the structure of example 2 due to the explosion of a bomb on the seventh floor for both the uncontrolled and the controlled structures. This structure has a relatively small aspect ratio of 1.75. It is observed that a structure with a smaller aspect ratio can be controlled more effectively.

8.5.3 Example 3:

Figure 8.14 shows the horizontal displacement at the top of the structure of example 3 due to the explosion of a bomb on the seventh floor for both the uncontrolled and the controlled structures. A comparison of Figures 8.13 and 8.14 shows that regular structures can be controlled more effectively than irregular structures.

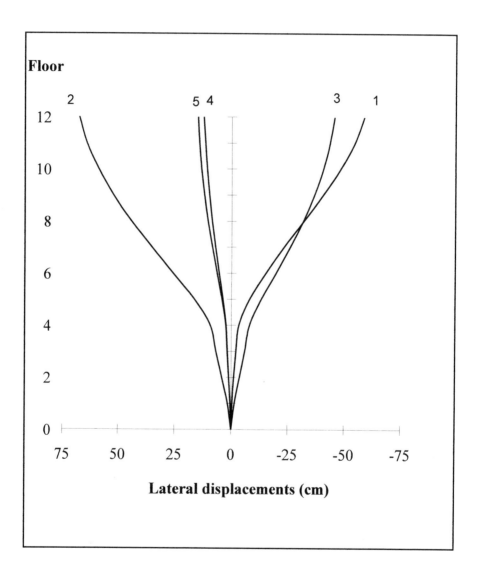

Figure 8.10: Lateral displacements of the uncontrolled structure of example 1 due to the explosion of a bomb
on the twelveth floor at time intervals of 0.0125 sec
(curve 1), 0.0375 sec (curve 2), 0.0625 sec (curve
3), 0.0875 sec (curve 4), and 0.1125 sec (curve 5).

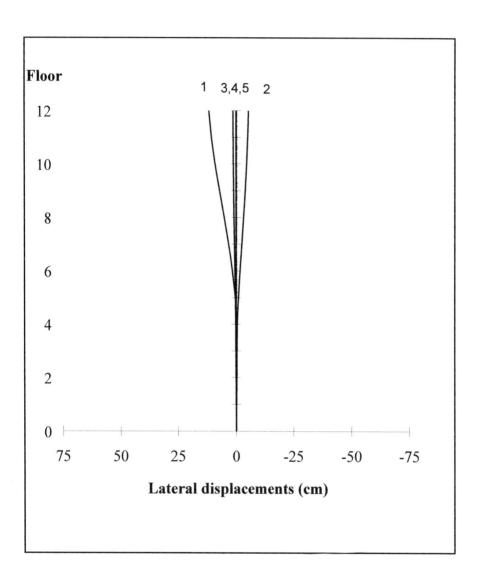

Figure 8.11: **Lateral displacements of the controlled structure of example 1 due to the explosion of a bomb on the twelveth floor at time intervals of 0.0125 sec (curve 1), 0.0375 sec (curve 2), 0.0625 sec (curve 3), 0.0875 sec (curve 4), and 0.1125 sec (curve 5).**

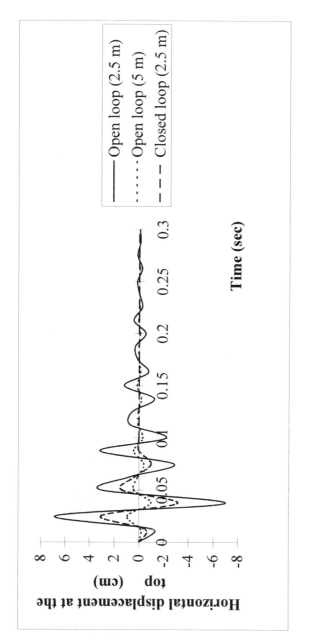

Figure 8.12: **The horizontal displacement at the top of the uncontrolled structure of example 1 due to the explosion of a bomb outside the building at distances 2.5 and 5.0 m. from the building and the controlled structure when a bomb is exploded at a distance 2.5 m. from the building.**

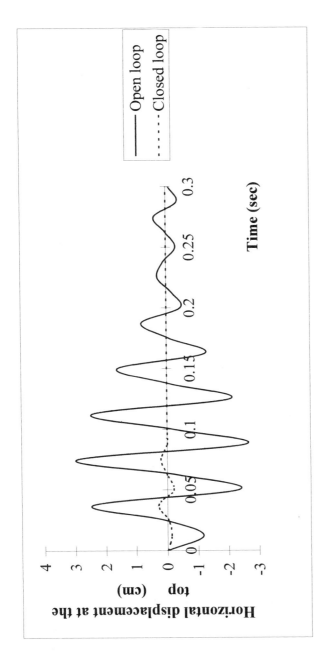

Figure 8.13: The horizontal displacement at the top of the structure of example 2 due to the explosion of a bomb on the seventh floor for both the uncontrolled and the controlled structures

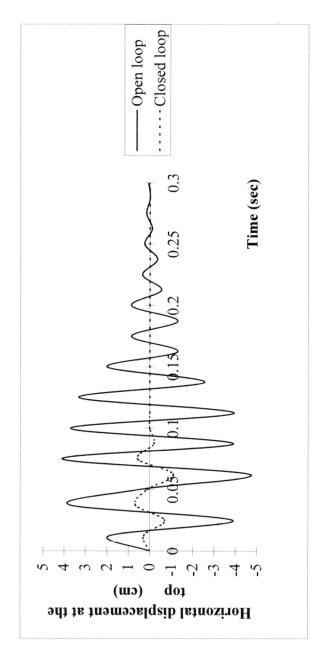

Figure 8.14: **The horizontal displacement at the top of the structure of example 3 due to the explosion of a bomb on the seventh floor for both the uncontrolled and the controlled structures.**

8.6 CONCLUSIONS

The general computational models and high-performance parallel algorithms developed by the authors have been used successfully for optimal control of large structures subjected to blast loadings. Both internal blast loading at different floor levels and external blast loading from outside the structure were considered. It is demonstrated that the response of a building structure can be reduced substantially to a fraction of the response of the uncontrolled structure. It should be noted that this work does not address catastrophic failure of structures due to large bombs. It is concerned with reducing the effects of resulting vibrations only. Other conclusions of this work are:

- For an explosive placed outside of the building, the effect increases drastically as its distance to the building decreases.
- Internal explosion in the upper floors causes the largest stress and interstory drift in a multistory building structure.
- For the multistory tier building structures when the internal bomb is placed on a floor roughly in the lower third height of the structure the maximum displacement and distress occur at the level of that floor. Otherwise, they occur at the top of the structure.
- Structures with smaller aspect ratios can be controlled more effectively.
- Regular structures can be controlled more effectively compared with irregular structures with setbacks or asymmetry in the plan.

CHAPTER 9

SIMULTANEOUS OPTIMIZATION OF CONTROL SYSTEM AND STRUCTURE

9.1 INTRODUCTION

Successful creation of adaptive or smart structures requires ingenious integration of four different technologies with structural engineering: control theory, optimization theory, sensor/actuator technology, and high-performance computing. The need for integration of control and optimization theories arises because of the desire to minimize both the required level of control forces and the weight of the structure. The formulation of such an integrated structural/control optimization problem is complex. Its solution for large structures with a few hundred members and more requires high-performance computing resources in terms of both memory and CPU. Thus, there is a need to develop efficient concurrent algorithms utilizing the unique architecture of multiprocessor supercomputers.

In Chapter 3, the formulation of the integrated structural/ control optimization problem was presented. In Chapters 4-6, the authors presented a general computational model for active control of large structures subjected to dynamic loading such as impact, earthquake, or wind loadings. *For a given structure* the response was minimized employing appropriately placed controllers. No attempt was made to optimize the weight of the structure.

When the response of a structure (displacement and stresses) is reduced below the code-specified value it is in general possible to reduce the weight of the structure at the cost of increasing the response as long as the response is within the code limits. In this chapter the goal is to minimize the weight of the structure as well as the required level of control forces. This is done by adding another external *layer* to the formulation that includes constraints on the stresses, displacements, complex parts of the closed loop eigenvalues, and their corresponding closed loop damping factors. The last two sets of constraints are used to ensure reduction of the response to the allowable limits.

9.2 PARALLEL ALGORITHMS

The approaches for high-performance computing on shared-memory supercomputers such as Cray YMP8E/8128 are vectorization of loops and multitasking approaches, such as microtasking and macrotasking, as explained in Chapter 2. We employ parallel stratagems with the objective of maximizing the speedup performance of parallel algorithms for the integrated structural and control optimization on shared memory multiprocessors using a combination of vector computations and multitasking approaches.

The outline of parallel algorithm for integrated structural and control optimization is presented in Table 9.1. This algorithm uses the parallel algorithms for computation of eigenvalues and eigenvectors for non-symmetric real matrices and solution of the Riccati equation presented in Chapters 4 and 5, respectively.

9.3 EXAMPLES

Three examples are presented: a one-span steel bridge structure (see Figure 6.1), a two-span continuous steel bridge structure (see Figure 6.2), and a steel multistory space moment resisting frame structure (see Figure 7.3).

Constraints are imposed on the imaginary part of the eigenvalues of the closed loop matrix, Eq. (3.26), and the corresponding damping factor, Eq. (3.27), so that the structure reaches the steady state in the least possible time, with minimum

Table 9.1: Outline of parallel algorithm for integrated structural and control optimization

Do sequentially
1. Set the number of processors (n_p).
2. Read in the input data and the starting design variables.
3. Set $1/\mu = 0.1$, iteration = 1 and operation = 1, where operation is a factor to indicate whether this step is in the analysis stage (operation = 1) or in the design stage (operation = 2).
4. Assemble the structure stiffness matrix.
   Do concurrently:
   > i- Calculate element stiffness matrices.   (***macrotasking***)
   > ii- Assemble element stiffness matrices into the structure stiffness matrix . (***microtasking with guarded region***)
5. Assemble total load vector.
   Do concurrently:
   Add the nodal forces into the structure load vector. (***microtasking***)
6. Apply boundary conditions.
   Do concurrently:
   > i- Update the structure stiffness matrix only if operation=1.
   > (***microtasking with guarded region***)
   > ii- Update total load vector.   (***microtasking***)
7. Solve the linear equations.
   Do concurrently:
   > i - Reduce the structure stiffness matrix only if operation =1.
   > (***vector computation***)
   > ii- Forward substitution.   (***microtasking***)
   > iii- Backward substitution. (***microtasking***)
8. If operation = 1,
   Do concurrently:

Table 9.1 – continued

Calculate the member forces and stresses. Go to step 9.
(*microtasking*)
If operation = 2, go to step 21.

9. Assemble the structure mass matrix
   Do concurrently:
   - i - Calculate the element mass matrices.   (*macrotasking*)
   - ii- Assemble the element mass matrices into the structure mass matrix. (*microtasking with guarded region*)

10. Calculate the eigenvalues and the eigenvectors for the homogeneous set of equations for the open-loop system by setting the right hand side of Eq. (3.1) to zero.
    (*Matrix multiplication, inner product, and normalization are done as in Tables 4.4 to 4.6*

11. De-couple equations of motion (Eqs. 3.3 to 3.6) and then reduce the second order differential equation to a first order equation. (Eqs. 3.7 to 3.11).
    (*Matrix multiplication is done concurrently, as in Table 4.4*).

12. Solve Riccati equation (Eq. 3.15) and find the Riccati matrix **P**.
    (*Use the algorithm in Table 5.1*).

13. Compute the gain matrix **G** (Eq. 3.14)
    (*Matrix multiplication is done concurrently, as in Table 4.4*).

14. Compute the eigenvalues for the closed-loop system (Eq. 3.20).
    (*Use the algorithms in Tables 4.1 to 4.6*).

15. Find the structural response and then the total displacement vector.
    (*Use the algorithm in Table 6.1*).

Table 9.1 – continued

16. Find member forces and stresses.
    Do concurrently:
        i - Calculate the element stiffness matrices. (***macrotasking***)
        ii- If operation = 1, calculate the total element forces and
           stresses; go to step 17. (***microtasking***)
        iii- If operation = 2, calculate the gradients of active
           displacement(s) according to Eq. (2.5); go to step 21.
           (***microtasking***)

17. **Do concurrently:**
    Calculate the objective function (W) using Eq. (3.22).
    (***microtasking***)
    If convergence, stop and print the results.
    Otherwise, go to step 18.

18. Set operation = 2.
    Do concurrently:
    Calculate the maximum displacement ratio. Go to step 19.
    (***microtasking***)

19. **Do concurrently:**
    Calculate the maximum stress ratio (stress_ratio).
    (**microtasking**)
    Go to step 20.

20. If iteration > 1 and the value of the objective function (W)
    is less than that of the previous iteration, set $1/\mu = (1/\mu)/2$.
    Go to step 21.

21. Apply unit load in the directions of the most active degrees of
    freedom one at a time; each time go to step 6-ii. When done
    with all the most active degrees of freedom go to step 22.

Table 9.1 – continued

22. **Do concurrently:**
    i - Calculate displacement sensitivities from Eq. (2.5).
    (***Matrix multiplication is done concurrently, as in Table 4.4***)
    ii - Calculate closed-loop eigenvalue sensitivities from Eq.
    (3.29). (***Matrix multiplication is done concurrently, as in Table 4.4***).
    iii- Calculate closed-loop damping factor sensitivities from
    Eq. (3.44). (***Matrix multiplication is done concurrently, as in Table 4.4***).
23. Calculate the new design variables for the next iteration as follows:
    Do concurrently:
    If stress_ratio > 1, modify the design variables by Eq. (2.13). (***microtasking***)
    Otherwise, modify the design variables by Eq. (3.48) or Eq. (2.12) (***microtasking***)
24. Check the closed-loop eigenvalue constraint in Eq. (3.26);
    If satisfied go to step 26.
    Otherwise, go to step 25.
25. **Do concurrently:**
    Modify the design variables by Eq. (3.51). (***microtasking***)
26. Check the closed-loop damping parameter constraint, Eq. (3.27); If satisfied go to step 28.
    Otherwise, go to step 27.
27. **Do concurrently:**
    Modify the design variables in Eq. (3.53). (***microtasking***)
28. Set iteration = iteration + 1. Set operation = 1.
29. Go to step 4.

number of oscillations, and maximum damping. A nonzero imaginary part ensures a damping factor of less than one which is necessary for damped oscillation of the structure (a damping factor of one results in a nonoscillatory motion that will vanish over a longer period of time). Experience shows a lower bound on the damping factor, ξ_j, in the range of 0.7 to 0.8 will minimize the deviation from the steady state response within the shortest time period. Similarly, for the lower bound on the imaginary part, $\overline{\omega}_j$, it was found that a value in the range of 0.75-1.0 times the real part of the smallest closed-loop eigenvalue, σ_j, (usually in the range of 1.0 to 2.0) will minimize the deviation from the steady state response within the shortest time period.

9.3.1   Example 1

This example is a single-span steel truss bridge described in section 6.4.1. Four controllers are placed along the members in every vertical plane passing through the joints over the middle half of the span. (see scheme B in Figure 6.1). The bridge is initially designed for AASHTO live load of H20 (AASHTO, 1993) and according to the American Institute of Steel Construction (AISC) Allowable Stress Design (ASD) specifications (AISC, 1989). Wide-flange shapes are selected for all the members of the bridge structure using A36 steel with yield stress of 248.2 MPa (36 ksi).

Then, three kinds of dynamic loadings a), b), and c) are considered, as described in Section 6.4. For the wind loading case c), a wind velocity of 160 kilometers per hour is used (q = 3.59 kN/m^2). Three other wind speeds of 80, 240, and 320 km/hr are also used to study the effect of the wind speed on the minimum weight controlled structure. The bridge structure is redesigned to find the minimum weight structure. The minimum weight of the uncontrolled structure with constraints on stresses and displacements is compared with minimum weight of the controlled structure with constraints on stresses, displacements, imaginary part of the closed loop eigenvalues and their corresponding damping factors. The limits on the imaginary parts of the smallest two closed loop eigenvalues, $\overline{\omega}_j$, in Eq. (3.26) are chosen as 1.75 and the limit on the corresponding damping factors, $\overline{\xi}_j$, in Eq. (3.27) is chosen as 0.7. This is based on the observation that for this structure, only the first two modes of vibration contribute to the response significantly.

9.3.2   Example 2

This is a two-span continuous truss steel bridge described in section 6.4.2. Four controllers are placed along the members in every vertical plane passing through the joints over the middle half of each span of the bridge (see scheme B in Figure 6.2). Loadings and constraints are the same as in Example 1.

9.3.3   Example 3

This example is a 12-story moment-resisting steel space frame described in section 7.3.3. The controllers are placed in the horizontal plane of each floor diaphragm alongside beams in two perpendicular directions (principal axes of the floor plan when there are two axes of symmetry in the plan) using two different schemes, A and B (see Figure 7.3). The loading on the structure consists of uniformly distributed dead and live loads of 2.88 kPa (60 psf) and 2.38 kPa (50 psf) (office building), respectively.

Wide-flange shapes from the AISC manual (AISC, 1989) are selected for all the members including bracings. In example 3a no bracing is used. In example 3b, cross bracings are used as identified in Figure 7.3a to study the effect of bracings on the response and stresses of the controlled structure.

Three kinds of dynamic loadings are considered a), b), and c), as explained in Section 7.3. For the wind loading cases b) and c), a wind velocity of 113 kilometers per hour is used ($q = 2.54$ kN/m^2). Two other wind speeds of 208 and 320 km/hr are also used to study the effect of the wind speed on the minimum weight controlled structure.

The minimum weight of the uncontrolled unbraced structure with constraints on stresses and displacements is compared with the minimum weight of the uncontrolled fully braced structure with constraints on stresses and displacements as well as the minimum weight of the controlled unbraced structure with constraints on the stresses, displacements, imaginary part of the closed loop eigenvalues and their corresponding damping factors. The interstory drift is limited to 0.004 times the story height. The limit on the imaginary part of the two smallest closed loop eigenvalues, $\overline{\omega}_j$, in Eq. (3.26) is chosen as 1.95 and the limit on the corresponding damping factors, ξ_j, in Eq. (3.27) is chosen as 0.725.

9.4 RESULTS
9.4.1 Example 1

Figure 9.1 shows the convergence histories for both uncontrolled and controlled structures subjected to combined impulsive

traffic loading a) and earthquake loading b). The minimum weight
obtained for the controlled structure is 655 kN compared with
minimum weight for the uncontrolled structure of 829 kN. The
minimum weight controlled structure is only 79% of the
corresponding minimum weight for the uncontrolled structure.
The level of required control forces in this case is in the range 3.74
kN to 69.0 kN.

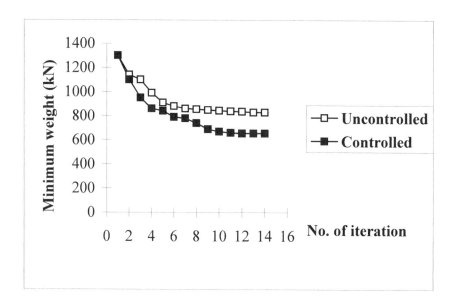

**Figure 9.1:   Convergence histories for example 1 subjected to
impulsive traffic and earthquake loadings**

Figure 9.2 shows the convergence histories for both
uncontrolled and controlled structures subjected to combined
impulsive traffic loading a) and wind loading c). The minimum
weight obtained for the controlled structure is 596 kN compared

with minimum weight for the uncontrolled structure of 716 kN. The minimum weight controlled structure is only 83% of the corresponding minimum weight for the uncontrolled structure.

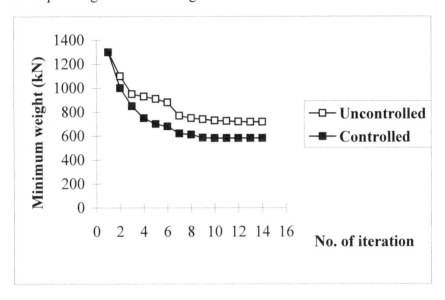

Figure 9.2: Convergence histories for example 1 subjected to impulsive traffic and wind loadings

Figure 9.3 shows the variation of the weights of the minimum weight uncontrolled and controlled structures subjected to combined traffic loading a) and wind loading c) using four different wind velocities of 80 km/h, 160 km/h, 240 km/h, and 320 km/h.

9.4.2 Example 2

Figure 9.4 shows the convergence histories for both uncontrolled and controlled structures subjected to combined impulsive traffic loading a) and earthquake loading b). The minimum weight obtained for the controlled structure is 729 kN compared with minimum weight for the uncontrolled structure of 947 kN. The minimum weight controlled structure is only 77% of the corresponding minimum weight for the uncontrolled structure. The level of required control forces in this case is in the range 2.73 kN to 57.0 kN.

Figure 9.5 shows the variation of the weights of the minimum weight uncontrolled and controlled structures subjected to combined traffic loading a) and wind loading c) using four different wind velocities of 80 km/h, 160 km/h, 240 km/h, and 320 km/h.

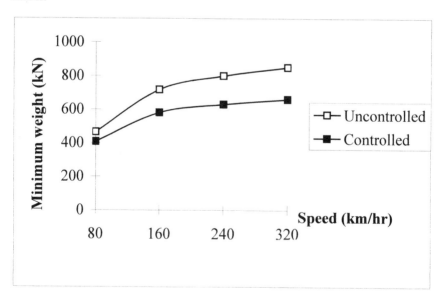

Figure 9.3: Minimum weight uncontrolled and controlled structure for example 1 subjected to impulsive traffic
and wind loadings using four different wind speeds

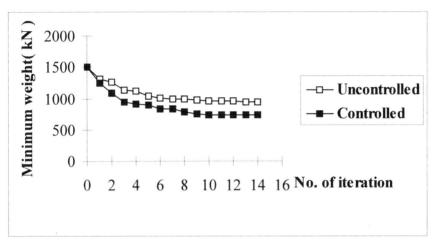

**Figure 9.4: Convergence histories for example 2 subjected to
impulsive traffic and earthquake loadings**

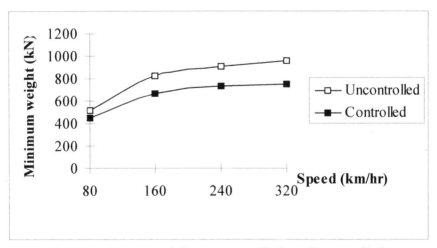

**Figure 9.5: Minimum weight uncontrolled and controlled
structure for example 2 subjected to impulsive
traffic loading and impulsive wind loadings using
four different wind speeds**

9.4.3 Example 3

Figure 9.6 shows the convergence histories for uncontrolled unbraced structure, uncontrolled braced structure, and controlled structure with two different schemes A and B for controllers subjected to earthquake loading b). The minimum weights obtained for the uncontrolled unbraced and braced structures are 678 kN and 613 kN, respectively. The minimum weight controlled structure is 542 kN using controllers scheme A and 548 kN using controllers scheme B.

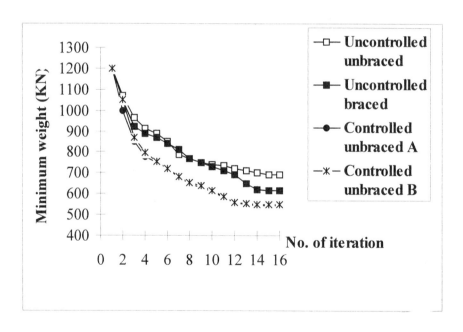

Figure 9.6: Convergence history for example 3 subjected to earthquake loading

The level of required control forces for scheme A is in the range 8.64 kN to 144.0 kN and for scheme B is in the range 9.24

kN to 154.0 kN. The maximum control forces are at the top of the structure. For scheme A, the level of control forces decreases gradually to 49% of the maximum force at the mid-height of the structure and then decreases rapidly to 6% of the maximum force at the first floor. For scheme B, the level of control forces decreases to 37% of the maximum force at the mid-height of the structure.

Figures 9.7 and 9.8 show variations of the weight of the minimum weight structure with the wind velocity for the case of symmetric wind loading b) and unsymmetric (twister) wind loading c), respectively.

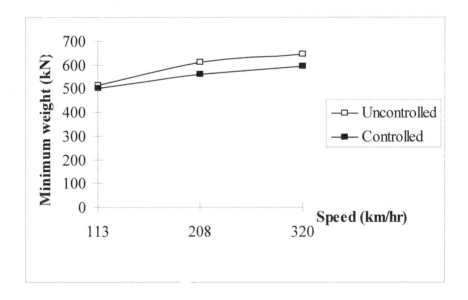

**Figure 9.7: Minimum weight uncontrolled and controlled struc-
tures for example 3 subjected symmetric wind
loading using three different wind speeds**

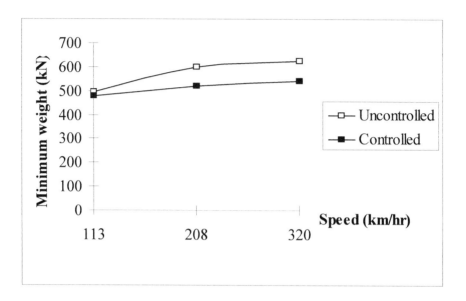

**Figure 9.8:   Minimum weight uncontrolled and controlled struc-
tures for example 3 subjected to unsymmetric
(twister) wind loading using three different wind
speeds**

9.5 MEMORY AND CPU REQUIREMENTS

The solution of the integrated structural/control optimization
problem for large-scale structures requires high-performance
computing resources and large memory. It is important for the
reader to have an appreciation of the size and complexity of the
problem being solved.

The algorithms developed in this work have been implemented
in C on the shared-memory supercomputer Cray YMP8E/8128
with 8 processors with a theoretical peak performance of 320
MFLOPS/processor (2.5 GFLOPS for 8 processors) and maximum
shared memory (RAM) of 1024MB (128Mwords). However, the

amount of RAM allocated to this work for processing time of more than 10 hrs was limited to 192MB. Mainframe computers can have a relatively large amount of memory. For example, the mainframe computer IBM 3090 can have a memory as large as 512MB. But, their processing power is only a fraction of supercomputers with vectorization and parallel processing capabilities.

Hwang (1993) reports that one processor of Cray YMP8E/8128 without vectorization is 3.1 times faster than IBM 3090 and one processor of Cray C90, introduced in 1995, is 9.2 times faster than IBM 3090. Our experience with Cray supercomputers indicates that vectorization increases the speedup in the range of 10 (for complicated problems with large dependencies) to 20 (for simple problems). Thus, taking into account the vectorization capability, the Cray YMP8E/8128 is at least 30 times faster than IBM 3090, even without using its parallel processing capability.

The largest example structure presented in this chapter (Example 3) has $N=408$ displacement degrees of freedom and $2N=816$ state variables. In each iteration, the solution of the resulting Riccati equation requires the solution of a complex eigenvalue problem for a matrix of size $4N \times 4N$ or 1632×1632. Subsequent to the solution of this eigenvalue problem, 816 sets of 816 complex linear equations must be solved. And this example needed 16 iterations to converge. Further, in each iteration, a complex eigenvalue problem of size 816×816 is solved four times (twice for the open loop system and twice for the closed loop system). The CPU time for *one iteration* of the integrated structural/control optimization problem using one processor of Cray YMP8E/8128 with vectorization was found to be 3 hours and 21 minutes with vectorization performance of 207 MFLOPS. A rough estimate of the time required to solve the same example on the mainframe IBM 3090 would be 101 hours per iteration and 67 days for the complete solution of the problem!

Now, we discuss the memory requirement. On the Cray YMP8E/8128, single and double precision real floating point variables require 8 and 16 bytes of memory, respectively. Single and double precision complex variables require 16 ad 32 bytes of memory, respectively. All other data types require 8 bytes of memory per variable which is twice the corresponding number on mainframe computers. The maximum amount of memory used in the solution of Example 3 was found to be 168 MB. This includes the total memory needed for matrices representing the structural and control systems as well as the matrices used for temporary storage. For example, the real matrix of size 4N x 4N or 1632 x 1632 for example 3, formed for the solution of the Riccati equation, requires 21.3 MB of memory. The resulting complex eigenvalues and corresponding complex eigenvectors of that matrix require 85.28 MB of memory. For the solution of each system of the 816 systems of 816 complex linear equations, 21.3 MB of memory is required.

The comparison of processing powers of mainframes and supercomputers with vectorization capability explains why high-performance multiprocessors are needed to solve the problem of integrated structural/control optimization for large structures. It has to be pointed out that we could have solved larger examples if more memory had been allocated for this work.

With the latest supercomputer technology, the present work can be applied to superhighrise building structures. For example, the latest Cray supercomputer, Cray C90 has 16 processors, maximum shared memory of 2.5 GB, and a peak theoretical processing speed of 16 GFLOPS. Using both vectorization and parallel processing, Cray C90 is roughly 14 times faster than the Cray YMP8E/8128.

9.6 FINAL COMMENTS

The primary conclusion is that through adroit use of active controllers the weight of the minimum weight structure can be reduced substantially. The result would be a substantially lighter structure for both bridge and building structures. It was also found that active controllers are more effective in the case of severe dynamic loadings such as earthquake loading in reducing the weight of the minimum weight structure. They are also more effective in reducing the weight of multispan continuous bridges than simply-supported single-span bridges. In the case of impulsive wind loadings, we observed that the effectiveness of active controllers increases with an increase in the speed of wind, as demonstrated by Figures 9.3, 9.5, 9.7, and 9.8.

With the rapid and continuous improvement of the actuator technology and increase in demand, their availability should increase and their price should decrease in coming years. Active controllers and the computational model and algorithms presented in this book give the future structural designers an effective means of reducing both the response of the structure and its weight. In some cases it may be desirable to reduce the response such as drift in high-rise and superhighrise building structures to a minimum. In other cases achieving the lightest structure while keeping the response within the code-specified limits may be desirable.

The computational model and algorithms developed in this work are admittedly complex and advanced and even perhaps a bit futuristic. But, they provide the theoretical foundation for design and construction of a new generation of high-technology adaptive/smart structures. We have already demonstrated their practicality by applying them to rather realistic example structures. The future is already here!

BIBLIOGRAPHY

1. AASHTO (1993), *Standard Specifications for Highway Bridges*, 15th Ed., American Association of State Highway and Transportation Officials, Washington.
2. Adeli, H. (1992a), *Supercomputers in Engineering Analysis*, Marcel Dekker, New York.
3. Adeli, H. (1992b), *Parallel Processing the Computational Mechanics*, Marcel Dekker, New York.
4. Adeli, H., Ed. (1994), *Advances in Design Optimization*, Chapman and Hall, London, UK.
5. Adeli, H. and Kamal, O. (1993), *Parallel Processing in Structural Engineering*, Elsevier Applied Science, London, UK.
6. Adeli, H. and Kumar, S. (1999), *Distributed Computer-Aided Engineering for Analysis, Design, and Visualization*, CRC Press, Boca Raton, FL.
7. Adeli, H. and Park, H. S. (1998), *Neurocomputing for Design Automation*, CRC Press, Boca Raton. FL.
8. Adeli, H. and Saleh, A. (1996), "Active Control of Large Structures," Proceedings of the International Conference on Strength, Durability, and Stability of Materials and Structures, Kaunas, Lithuania, September 18-20, pp. 9-14.

9. Adeli, H. and Saleh, A. (1997a), "Optimal Control of Adaptive/Smart Bridge Structures," *Journal of Structural Engineering, ASCE*, Vol. 123, No. 2, pp. 218-226.

10. Adeli, H. and Salah, A. (1997b), "Active Control of Large Adaptive Structures Subjected to Dynamic Loadings," Proceedings of the 8th International Symposium on Interaction of the Effects of Munitions with Structures," McLean, VA, April 22-25, published by Defense Nuclear Agency, Albuquerque, NM, pp. 981-987.

11. Adeli, H. and Saleh, A. (1998), "Integrated Structural/Control Optimization of Large Adaptive/Smart Structures," *International Journal of Solids and Structures*, Vol. 35, Nos. 28-29, pp. 3815-3830.

12. Adeli, H. and Soegiarso, R. (1999), *High-Performance Computing in Structural Engineering*, CRC Press, Boca Raton, FL.

13. AISC (1989), *Manual of Steel Construction-Allowable Stress Design*, 8th Ed., American Institute of Steel Construction, Chicago.

14. Armstrong, E. S. (1978), "ORACLS - A system for Linear Quadratic Gaussian Control Law Design," NASA Technical Paper 1106, Washington, D.C.

15. Armstrong, E. S. (1980), "ORACLS - A Design system for Linear Multivariable Control," Marcel Dekker, Inc., NewYork, NY.

16. Arnold, W. F. and Laub, A. J. (1984), "Generalized Eigenproblem Algorithms and Software for Algebraic Riccati Equations," *Proceedings of the IEEE*, Vol. 72, No. 12, pp. 1746-1754.

17. Bathe, K. J. (1982), *Finite Element Procedures in Engineering Analysis*, Prentice Hall, Englewood Cliffs, New Jersey.
18. Britton, R. R. (1983), *The Effects of Decoupling on Rock Breakage*, M.S. Thesis, the Ohio State University.
19. Britton, R. R., Konya, C. J., and Skidmore, D. R. (1984), "Primary Mechanism for Breaking Rock with Explosives," *Proceedings of the Twenty-fifth Symposium on Rock Mechanics*, Northwestern University, Evanston, Illinois, June 25-27, pp. 942-949.
20. Brockenbrough, R. L. and Merritt, F. S. (1994), *Structural Steel Designer's Handbook*, 2nd Ed., McGraw-Hill, Inc., N.Y.
21. Byers, R. (1987), "Solving the Algebraic Riccati Equation with the Matrix Sign Function," *Linear Algebra and Its Applications*, Vol. 85, pp. 267-279.
22. Casti, J. L. (1987), *Linear Dynamical Systems*, Academic Press, Inc., Orlando, Florida.
23. Chopra, A. K. (1995), *Dynamics of Structures: Theory and Applications to Earthquake Engineering*, Prentice Hall, Englewood Cliffs, NJ.
24. Cole, R. H. (1948), *Underwater Explosions*, Princeton University Press, Princeton, NJ.
25. CRAY (1987), *Cray Y-MP Multitasking Programmer's Reference Manual*, SR-0222, Cray Research Inc., 1987.
26. CRAY (1990), *Cray Standard C Programmer's References Manual*, SR-2074 3.0, Cray Research Inc., 1990.
27. D'Azzo, J. J. and Houpis (1989), *Linear control system analysis and design*, 2nd Ed., McGraw-Hill, New York.
28. Dongarra, J. J. and Sidani, M. (1993), "A Parallel Algorithm For The Nonsymmetric Eigenvalue Problem," *SIAM, Journal on Scientific Computing*, Vol. 14, No. 3, pp. 542-569.

29. Fental, M. J. (1996), "Design for Blast Protection," *Civil Engineering, ASCE*, September Issue: 3A-5A.
30. Franklin, J. (1968), *Matrix Theory*, Prentice-Hall, Inc., Englewood Cliffs, New Jersey.
31. Friedland, B. (1986), *Control System Design: An Introduction to State Space Methods*, McGraw Hill, Inc., New York, NY.
32. Furuya, H. and Haftka, R. T. (1995), "Placing Actuators on Space Structures by Genetic Algorithms and Effectiveness Indices," *Structural Optimization*, Vol. 9, No. 2, pp. 69-75.
33. Garbow, J. M., Boyle, J. M., Dongarra, J. J., and Moler, C. B. (1977), *Matrix Eigensystem Routines - EISPACK Guide Extension, Lecture Notes in Computer Science*, Vol. 51, Springer-Verlag, Berlin.
34. Gardiner, J. D. (1997), "A Stabilized Matrix Sign Function Algorithm for Solving Algebraic Riccati Equations," *SIAM, Journal on Scientific Computing*.
35. Gardiner, J. D. and Laub, A. J. (1986), "A Generalization of the Matrix-Sign-Function Solution for the Algebraic Riccati Equations," *International Journal of Control*, Vol. 44, No. 3, pp. 823-832.
36. Golub, G. H. and Van Loan, C. F. (1989), *Matrix Computations*, 2nd Ed., John Hopkins University Press, Baltimore.
37. Hewer, G. A. (1971), "An Iterative Technique for the Computation of the Steady State Gains for the Discrete Optimal Regulator," *IEEE Transactions on Automatic Control*, Vol. AC-16, No. 4, pp. 382-383.
38. Housner, G. W., Soong, T. T., and Marsi, S. F. (1996), "Second Generation of the Active Structural Control in Civil

Engineering," *Microcomputers in Civil Engineering*, Vol. 11, No. 5, pp. 289-296.

39. Hwang, K. (1993), *Advanced Computer Architecture, Parallelism, Scalability, Programmability*, McGraw Hill, New York.

40. Johnson, C. D. (1971), "Accommodation of External Disturbances in Linear Regulators and Servomechanism Problems," *IEEE Transactions on Automatic Control*, Vol. AC-16, No. 6, pp. 635-643.

41. Khan, M. R., Willmert, K. D., and Thornton, W. A. (1979), "An Optimality Criterion Method for Large-Scale Structures," *AIAA Journal*, Vol. 17, No. 7, pp. 753-761.

42. Khot, N. S. (1994), "Optimization of Controlled Structures," in Adeli, H., Ed., *Advances in Design Optimization*, Chapman and Hall, London, pp. 266-296.

43. Khot, N. S., Berke, L., and Venkayya, V. B. (1978), "Comparison of optimality Criteria Algorithms for Minimum Weight Design of Structures," *AIAA Journal*, Vol. 17, No. 2, pp. 182-190.

44. Khot, N. S., Eastep, F. E., and Venkayya, V. B. (1985a), "Optimal Structural Modifications to Enhance the Optimal Active Vibration Control of Large Flexible Structures," *AIAA Paper* 85-0627, Orlando, Florida.

45. Khot, N. S., Eastep, F. E., and Venkayya, V. B. (1985b), "Simultaneous Optimal Structural/Control Modifications to Enhance the Vibration Control of Large Flexible Structures," *AIAA Paper* 85-1925, Snowmass, Colorado.

46. Kleinman, D. L. (1968), "On an Iterative Technique for Riccati Equation Computations," *IEEE Transactions on Automatic Control*, Vol. AC-13, No. 1, pp. 114-115.

47. Kobori, T. (1996), "Future Directions on Research and Development of Seismic-Response-Controlled Structures," *Microcomputers in Civil Engineering*, Vol. 11, No. 5, 297-304.
48. Konya, C. J. and Walter, E. J. (1990), *Surface Blast Design*, Prentice Hall, Englewood Cliffs, New Jersey.
49. Laub, A. J. (1979), "A Schur Method for Solving Algebraic Riccati Equations," *IEEE Transactions on Automatic Control*, Vol. AC-24, No. 6, pp. 913-921.
50. Lindmann, C. (1998), *Performance Modeling with Deterministic and Stochastic Petri Nets*, John Wiley & Sons, New York.
51. Liu, S. C., Eeri, M., Lagorio, H. J., and Chong, K. P. (1991), "Status of U.S. Research on Structural Control Systems," *Earthquake Spectra*, Vol. 7, No. 4, pp. 543-550.
52. Marks II, R. J. (1993), *Advanced Topics in Shannon Sampling and Interpolation Theory*, Springer-Verlag, Berlin.
53. Meirovich, L. (1980), *Computational Methods in Structural Dynamics*, Sijthoff & Noordhoff, Alphen ann den Rijn, The Nethurlands.
54. Meirovich, L. (1985), *Introduction to Dynamics and Control*, John Wiley & Sons, New York.
55. Meirovich, L. (1990), *Dynamics and Control of Structures*, John Wiley & Sons, New York.
56. Mohammed, J. L. and Walsh, J. L. (1986), *Numerical Algorithms*, Clarendum Press, Oxford, UK.
57. Morse S. and Wonham, W. H. (1971), "Status of Noninteracting control," *IEEE Transactions on Automatic Control*, Vol. AC-16, No. 6, pp. 568-581.

58. Paige, C. and Loan, C. V. (1981), "A Schur Decomposition for Hamiltonian Matrices," *Linear Algebra and Its Applications*, Vol. 50, pp. 11-32.

59. Potter, J. E. (1966), "Matrix Quadratic Solutions," *SIAM Journal of Applied Mathematics*, Vol. 14, No. 3, pp. 496-501.

60. Rhodes, L. B. (1971), "A Tutorial Introduction to Estimation and Filtering," *IEEE Transactions on Automatic Control*, Vol. AC-16, No. 6, pp. 688-706.

61. Rutishauser, H. (1990), *Lectures on Numerical Mathematics*, Birkhauser, Boston.

62. Saleh, A. and Adeli, H. (1992), "Multitasking Algorithm for Optimization of Space Structures," Proceedings of SUPERCOMPUTING 92, Minneapolis, MI, November, 16-20.

63. Saleh, A. and Adeli, H. (1994a), "Microtasking, Macrotasking, and Autotasking for Structural Optimization," *Journal of Aerospace Engineering, ASCE*, Vol. 7, No. 2, pp. 156-174.

64. Saleh, A. and Adeli, H. (1994b), "Parallel Algorithms for Integrated structural/control Optimization," *Journal of Aerospace Engineering, ASCE*, Vol. 7, No. 3, pp. 297-314.

65. Saleh, A. and Adeli, H. (1996), "Parallel Eigenvalue Algorithms for Large Scale Control-Optimization Problems," *Journal of Aerospace Engineering, ASCE,* Vol. 9, No. 3, pp. 70-79.

66. Saleh, A. and Adeli, H. (1997a), "Robust Parallel Algorithms for Solution of the Riccati Equation," *Journal of Aerospace Engineering, ASCE*, Vol. 10, No. 3, pp. 126-133.

67. Saleh, A. and Adeli, H. (1997b), " Large Adaptive Structures," in Adeli, H., Intelligent Information Systems, IEEE Computer Society, Los Alamitos, CA, pp. 564-568.

68. Saleh, A. and Adeli, H. (1998a), "Optimal Control of Adaptive/Smart Multistory Building Structures," *Computer-Aided Civil and Infrastructure Engineering,* Vol. 13, No. 6, pp. 389-403.

69. Saleh, A. and Adeli, H. (1998b), "Optimal Control of Adaptive Building Structures Under Blast Loading," *Mechatronics,* Vol. 8, No. 8, pp. 821-844.

70. Sandell, N. R., Jr. (1974), "On Newton's Method for Riccati Equation Solution," *IEEE Transactions on Automatic Control,* Vol. AC-19, No. 3, June, pp. 254-255.

71. Stewart, G. W. (1973), *Introduction to Matrix Computations,* Academic Press, Orlando, Florida.

72. Stewart, G. W., Moler, C. B., Dongarra, J. J., and Bunch, J. R. (1979), *Linpack User's Guide,* Society of Industrial and Applied Mathematics, SIAM, Philadelphia, PA.

73. Stubbs, N. and Park, S. (1996), "Optimal Sensor Placement for Mode Shapes via Shannon's Sampling Theorem," *Microcomputers in Civil Engineering,* Vol. 11, No. 6, pp. 411-419.

74. UBC (1994), *Uniform Building Code,* Vol. 2- *Structural Engineering Design Provisions,* International Conference of Building Officials, Whitter, CA.

75. Walsh, J. L. (1967), *Numerical Analysis An Introduction,* Thompson Book Company, Washington, D. C.

76. Wilkinson, J. H. (1963), *Rounding Errors in Algebraic Processes,* Series in Automatic Computation, Prentice-Hall, Englewood Cliffs, New Jersey.

77. Wilkinson, J. H. (1965), *The Algebraic Eigenvalue Problem,* Oxford University Press, London.

78. Wilkinson, J. H. and Reinsch, C. (1965), *Handbook for Automatic Computation: Vol II - Linear Algebra*, Springer-Verlag, New York.

SUBJECT INDEX

-